KITSON MEYER
ARTICULATED LOCOMOTIVES

THE DEFINITIVE HISTORY   –   DONALD BINNS

A LOCOMOTIVES INTERNATIONAL SPECIAL NUMBER

The last surviving Kitson Meyer 0–6+6–0T– Taltal Railway No. 59, leaving Taltal.   *A. E. Durrant.*

Designed by Donald Binns. Copyright 1993. Locomotive International/D. Binns.
Published by Locomotives International. 50 Long Meadow, Skipton, North Yorkshire, England, BD23 1BW.
Typesetting by Hanson Typesetting Services Ltd., 68 Haworth Road, Cross Roads, Keighley, West Yorkshire
Printed by Clifford Ward & Co. Ltd., Bridlington, East Yorkshire.

1

THE KITSON MEYER DESIGN AND THE DIFFERENT STAGES IN ITS DEVELOPMENT

In 1894 Robert Stirling ordered some Meyer locomotives from Kitson & Company (Leeds) for his 3ft 6in gauge Anglo-Chilian Nitrate & Railway Company, a South American nitrate carrier serving the districts around Toco. A modified form of the Meyer was elaborated—hence the name Meyer-Kitson (in this work referred to as the Kitson-Meyer). Just how these alterations were devised to produce the distinctive Kitson Meyer form of articulation is rather uncertain, but Robert Stirling—Locomotive Superintendent of the FCTT (and son of Patrick Stirling of the Great Northern Railway, England) is known to have been extremely keen on the idea of articulation. Stirling had knowledge of the various different forms of articulated locomotives in use and in particular with the Meyer type developed in Europe.

For reasons which will be obvious later on, it seems reasonable to assume that Stirling knew of and had probably inspected the 0−6+6−0T built by the Baldwin Locomotive Works in 1892 for the Sinnemahoning Valley Railroad. Obviously dissatisfied with the two Fairlie articulateds on the FCTT, Stirling suggested to Kitson & Company a locomotive "with the characteristics of a superstructure on two steam driven bogies". Kitson & Company supposed that Stirling had in mind the "propositions of Mr. J. J. Meyer of France". Stirling however, had other ideas, and from his accumulated knowledge developed his own articulated type for which Kitson & Company subsequently claimed most of the credit. The new design was virtually a copy of the 1892 Baldwin and it seems certain that Stirling must have examined this either under construction or in service. Kitson & Company were of course familiar with the French and Belgian Meyer locomotive types but only had general dimensions of the 1892 Baldwin modification. Although Kitson & Company gave little credit the so-called Kitson Meyer articulated locomotive seems to have been the brain child of Robert Stirling and without him it most certainly would never have existed. The prototype Kitson Meyer emerged in 1894 and looked embarassingly like the Baldwin locomotive. Kitson & Company of course denied knowledge of this and the position was supposedly clarified in 1920 when Lt. Col. Edwin Kitson Clark said that "in 1894 Mr. Robert Stirling of the Anglo Chilian Nitrate & Railway Company, having to deal with a line 17 miles long of 1 in 25 gradients, 75% of which was combined with curves of 181ft radius, suggested to Messrs. Kitson, a design on the lines of that made by Meyer. The principals of the firm had already noted the type on the Continental railways and it has since developed in many various designs". To this was added a footnote—"It was subsequently found that an engine of this type had been made two years before by the Baldwin Company. It was compounded on the Vauclain system, but beyond general dimensions, no particulars were available".

The prototype Kitson Meyer was the first of 13 to be supplied to that Company. It had two 6-coupled steam bogies (0−6+6−0T), each driven from its own pair of 14in × 18in outside cylinders mounted at the rear of each unit. The cylinders were provided with special covers which allowed them to be dismantled without interfering with the side tanks above. The superstructure comprising boiler, cab, fuel and water compartments were laid out in conventional positions on a girder frame supported on the steam bogies. Pivots were built into the superstructure working in sockets integral with the steam bogies and both pivots and sockets were of ample size being located as close to the centre of the adhesive wheelbase as possible. The driving wheels were 2ft 10¾in in diameter and the rigid wheelbase of each group was a mere 6ft 2½in with a total wheelbase of 25ft 6½in. With an adhesive weight of 55 tons 7 cwt Stirling's articulated was an extremely flexible and compact machine with a hefty 27,600lb tractive effort at 85% of the 160lb boiler pressure. The girder mounted superstructure was prevented from rolling excessively by plates concentric with each pivot and from pitching by slides positioned at the end of the steam bogies. In true Meyer designs the leading steam bogie supplemented the traction of the rear unit by means of a long draw-bar which grasped the pivot casting. In the Stirling/Kitson & Company design this feature was abandoned and the pull of the two bogies was co-ordinated steadily by the connection of the two pivots built into the girder superstructure. The design produced a compact locomotive with only the firebox between the steam bogies and this could therefore be of ample size and was in no way restricted by the positioning of the bogies. In the early designs all the locomotive weight was available for adhesion but as the Kitson Meyer developed, leading and trailing pony trucks were incorporated. According to advertising literature of the day the boiler was easy to wash out and both firebox and ashpan could be to any required size with the ashes quickly emptied. At first, allowance was made for adjustment of the relative loads on the wheels by transferring part of the steam bogie weight to the superstructure (or vice versa), using adjusting bolts in the end slides. After one incident resulting from incorrect adjustment, Kitson & Company dispensed with this feature and from that time the whole weight of the superstructure was transferred to the pivot sockets. As in the original Meyer design (but unlike the Baldwin locomotive) the Kitson Meyer had its drawgear mounted on the steam bogies. Because of relative movement of the component parts and the effects of expansion under heat, some form of relief was necessary for both steam and exhaust pipes and this was allowed for by pipes sliding in glands and ball and socket joints, the centre of the ball coinciding with the centre of the spherical pivot casting.

In early Kitson Meyer designs, exhaust from the rear steam

The "Edward T. Johnson" was based on the Meyer principle of articulation and Baldwin Works No. 12526/1892 carried road No. 3 of the Sinnemahoning Valley Railroad—then part of the Pennsylvanian-based Goodyear logging empire. This Vauclain compound 0−6+6−0T became No. 103 on the Buffalo & Susquehanna Railroad in 1893 when the SVRR lines formed the basis for the new company. It is believed this Meyer was returned to the Baldwin Locomotive Works for disposal, but no further information is available. The European Meyer designs had their cylinders at the inner ends of each unit whilst the Baldwin was the fore-runner of the Kitson Meyer type 1 in which the cylinders were at the rear of each unit. D. Binns collection.

One of the first three Kitson Meyer 0−6+6−0T built by Kitson & Co. in 1894 (Works No's 3532-4) for the 3ft 6in gauge Anglo Chilian Nitrate & Railway Company. The small diameter rear chimney is interesting—one if not both of the other 1894 units had a normal diameter rear chimney. Note the similarity to the "Edward T. Johnson. When photographed, the front steps had not been fitted and would in all probability have been supplied separately for fitting by the purchasing company. D. Binns collection.

unit was directed through a rear chimney on the bunker behind the cab, the exhaust passing through piping in the water tanks slightly pre-heating the feed water. In later designs rear steam bogie exhaust was routed back to the smoke-box where it was exhausted in the usual way. In later years the Kitson Meyer was an excellent design capable of speeds up to 50mph and would undoubtedly have been constructed in larger numbers had it not been for the Beyer Garratt articulated. It is ironical to record that in 1907 H. W. Garratt had called on Kitson & Company, Airedale Foundry, Leeds, to see if they would undertake construction of his patent articulated. Naturally at that time Kitson & Company were not interested in another engineers "damned improvements". The Kitson Meyer was never extended to the limit of its possibilities and such a locomotive would have been a serious competitor for the Beyer-Garratt market. Indeed the later Kitson Meyer locomotives were superb machines ranking amongst the finest articulated power ever built.

KITSON & CO. RECORD OF CONSTRUCTION – KITSON MEYER ARTICULATED LOCOMOTIVES

In all, 78 units have been accounted for which carry Kitson & Co. works numbers—these include the pair of 2–8+8–2T erected by Robert Stephenson & Hawthorn following cessation of locomotive construction by Kitson & Co., although all parts for these were produced by Kitson & Co.

Works No's 3532, 3533, 3534 of 1894. 3 locos 3ft 6in gauge for the Anglo Chilian Nitrate & Railway Co. 0–6+6–0T.

Works No. 3604 of 1895. One loco 3ft 6in gauge for the Anglo Chilian Nitrate & Railway Co. 0–6+6–0T.

Works No. 4197 of 1903. One loco 3ft 6in gauge for the Cape Government Railways. 0–6+6–0 with tender.

Works No's 4240, 4241 of 1903. Two locos 3ft 6in gauge for the Rhodesia Railways. 0–6+6–0 with tender.

Works No's 4252, 4253, 4254 of 1904. Three locos 4ft 8½in gauge for the Jamaica Government Railways. 0–6+6–0T.

Works No. 4262 of 1904. One loco 3ft 6in gauge for the Central South African Railways. 0–6+6–0 with tender.

Works No. 4288 of 1904. One loco 3ft 6in gauge for the Taltal Railway. 0–6+6–0T.

Works No's 4432, 4433, 4434 of 1906. Three locos 3ft 6in gauge for the Taltal Railway. 0–6+6–0T.

Works No. 4488 of 1907. One loco 1 metre gauge for the Chilian Transandine Railway. 0–8+6–0 Rack/Adhesion Tank.

Works No's 4504, 4505, 4506, 4512, 4513, 4514 of 1907. Six locos 3ft 6in gauge for the Taltal Railway. 0–6+6–0T.

Works No. 4534 of 1908. One loco 2ft 6in gauge for the Antofagasta Railway. 2–6+6–4T.

Works No. 4568 of 1908. One loco 1 metre gauge for the Leopoldina Railway. 2–6+6–4T.

Works No's 4580, 4581, 4582 of 1908. Three locos for the 5ft 5¹³⁄₁₆in gauge Great Southern of Spain Railway. 2–8+8–0T.

Works No. 4598 of 1908. One loco for the 1 metre gauge Chilian Transandine Railway. 0–8+6–0 Rack/Adhesion Tank.

Works No's 4653, 4654, 4655, 4656 of 1909. Four locos 3ft 6in gauge for the Anglo Chilian Nitrate & Railway Co. 0–6+6–0T.

Works No. 4664 of 1909. One loco for the 1 metre gauge Chilian Transandine Railway. 0–8+6–0 Rack/Adhesion Tank.

Works No's 4669, 4670 of 1909. Two locos for the 1 metre gauge Argentine Transandine Railway. 0–8+6–0 Rack/Adhesion Tank.

Works No's 4671, 4672, 4673 of 1909. Three locos 3ft 0in gauge for the Colombian National Railways. 0–6+6–0T.

Works No. 4674 of 1909. One loco 1 metre gauge for the Argentine Transandine Railway. 0–8+6–0 Rack/Adhesion Tank.

Works No's 4735, 4736 of 1910. Two locos 3ft 6in gauge for the Anglo Chilian Nitrate & Railway Co. 2–6+6–2T.

Works No. 4841 of 1911. One loco for the 2ft 6in gauge Antofagasta Railway. 2–6+6–2T.

Works No. 4842 of 1911. One loco 1 metre gauge for the Argentine Transandine Railway. 0–8+6–0 Rack/Adhesion Tank.

Works No's 4853, 4854 of 1912. Two locos 3ft 6in gauge for the Anglo Chilian Nitrate & Railway Co. 2–6+6–2T.

Works No's 4882, 4883 of 1912. Two locos 1 metre gauge for the Argentine Transandine Railway. 0–8+6–0 Rack/Adhesion Tank.

Works No's 4915, 4916 of 1912. Two locos 3ft 0in gauge for the Colombian National Railways. 0–6+6–0T.

Works No's 4972, 4973, 4974, 4975 of 1913. Four locos 3ft 6in gauge for for the Manila Railway. 2–6+6–2T.

Works No's 5039, 5040 of 1914. Two locos 3ft 0in gauge for the Colombian National Railways. 0–6+6–0T.

Works No's 5064, 5065 of 1914. Two locos 3ft 0in gauge for the Colombian National Railways. 0–6+6–0T.

Works No's 5176, 5177, 5178 of 1918. Three locos 3ft 0in gauge for the Colombian National Railways. 0–6+6–0T.

Works No's 5274, 5275, 5276 of 1920. Three locos 3ft 0in gauge for the Colombian National Railways. 0–6+6–0T.

Works No's 5322, 5323, 5324, 5325 of 1921. Four locos 3ft 0in gauge for the Colombian National Railways. 0–6+6–0T.

Works No's 5400, 5401, 5402, 5403 of 1927. Four locos 3ft 0in gauge for the Colombian National Railways. 2–6+6–2T.

Works No's 5413, 5414 of 1928. Two locos 2ft 6in gauge for the North Western Railway of India (Kalka–Simla Railway. 2–6–2+2–6–2T.

Works No's 5416, 5417 of 1928. Two locos 3ft 0in gauge for the Cundinamarca Railway. 2–6+6–2T.

Works No. 5431 of 1929. One loco 3ft 0in gauge for the Cundinamarca Railway. 2–6+6–2T.

Works No's 5471, 5472 of 1935. Two locos 3ft 0in gauge for the Colombian National Railways. 2–8+8–2T.

Total Construction 78 units.

In addition there were a further 9 examples built by makers other than Kitson & Co. for use in South America.

Kerr Stuart & Co. Works No. 816 of 1903. One Kitson Meyer type 0–6+6–0T. 3ft 6in gauge for the Anglo Chilian Nitrate & Railway Co.

Beyer Peacock & Co. Works No's 5617–22 of 1913. Six Meyer type 0–6–2+0–6–2 with tenders for the 1 metre gauge Bolivia Railway.

Yorkshire Engine Co. Works No's 940 and 941 of 1912. Two Meyer type 0–6+6–0T 4ft 8½in gauge for the Nitrate Railways of Chile.

The Kitson design passed through various stages of development and the principal types may be sub-divided as follows:

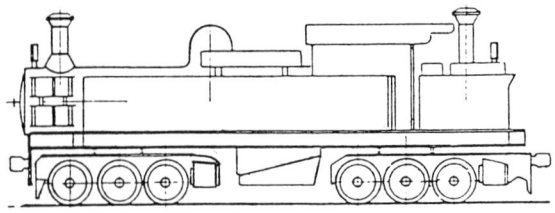

1. Simple Tank Locomotives with the cylinders at the rear of each steam bogie. Most (if not all) locomotives in this category had full length side tanks and all the early production had rear chimneys. Type 1 was built as follows:

Works No.	Year	Gauge	Railway	Wheels	Rear Stack	Side Tanks
3532	1894	3ft 6in	ACN&R	0−6+6−0T	Yes	Full
3533	,,	,,	,,	,,	,,	,,
3534	,,	,,	,,	,,	,,	,,
3604	1895	,,	,,	,,	,,	,,
4252	1904	4ft 8½in	JGR	,,	,,	,,
4253	,,	,,	,,	,,	,,	,,
4254	,,	,,	,,	,,	,,	,,
4288	,,	3ft 6in	Taltal	,,	,,	,,
4432	1906	,,	,,	,,	,,	,,
4433	,,	,,	,,	,,	,,	,,
4434	,,	,,	,,	,,	,,	,,
4504	1907	,,	,,	,,	,,	,,
4505	,,	,,	,,	,,	,,	,,
4506	,,	,,	,,	,,	,,	,,
4512	,,	,,	,,	,,	,,	,,
4513	,,	,,	,,	,,	,,	,,
4514	,,	,,	,,	,,	,,	,,
4653	1909	,,	ACN&R	,,		,,
4654	,,	,,	,,	,,		,,
4655	,,	,,	,,	,,		,,
4656	,,	,,	,,	,,		,,
4671	,,	3ft 0in	CNR	,,	,,	,,
4672	,,	,,	,,	,,	,,	,,
4673	,,	,,	,,	,,	,,	,,
4915	1912	,,	,,	,,		
4916	,,	,,	,,	,,		
5039	1914	,,	,,	,,		
5040	,,	,,	,,	,,		
5064	,,	,,	,,	,,		
5065	,,	,,	,,	,,		
5176	1918	,,	,,	,,		
5177	,,	,,	,,	,,		
5178	,,	,,	,,	,,		
5274	1920	,,	,,	,,	No	Full
5275	,,	,,	,,	,,	,,	,,
5276	,,	,,	,,	,,	,,	,,
5322	1921	,,	,,	,,		
5323	,,	,,	,,	,,		
5324	,,	,,	,,	,,		
5325	,,	,,	,,	,,		

It is thought that all the above had full length water tanks but where no entry is shown in the above list it has not been possible to confirm. However, it seems reasonably safe to state that probably all had full length side tanks.

Similarly it has not been possible to prove where the change-over came from having rear chimneys to not having them. In all probability the 1914 locomotives would have rear chimneys and the 1918 locomotives would not.

Total construction Type 1 0−6+6−0T 40 units (or a little over half the total production).

In addition one identical unit was built by Kerr Stuart & Co.

Works No.	Year	Gauge	Railway	Wheels	Rear Stack	Side Tanks
816	1903	3ft 6in	ACN&R	0−6+6−0T	Yes	Full

Also two units 0−6+6−0T were built by the Yorkshire Engine Co., Sheffield.

Works No.	Year	Gauge	Railway	Wheels	Rear Stack	Side Tanks
940	1912	4ft 8½in	NR	0−6+6−0T	Yes	¾ approx.
941	,,	,,	,,	,,	,,	,,

The Second type of Kitson Meyer to be introduced was derived from type 1 and in this case was devoid of side tanks but was provided with a separate tender. The cylinders were still at the rear of each steam bogie and in this work these are recorded as Type 2. These were the least successful of the various types since a coal carrying tender attached to the locomotive was far from ideal. Had these been oil-burners they would probably have been much more successful.

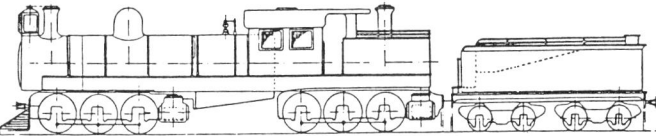

Works No.	Year	Gauge	Railway	Wheels	Rear Stack	Side Tanks
4197	1903	3ft 6in	CGR	0−6+6−0	Yes	None
4240	,,	,,	RR	,,	,,	,,
4241	,,	,,	,,	,,	,,	,,
4262	1904	,,	CSAR	,,	,,	,,

Total Construction Type 2 0−6+6−0 with separate 8-wheel tenders 4 units.

In addition six 0−6−2+0−6−2 with separate tenders were built by Beyer Peacock & Co. Ltd.

Works No.	Year	Gauge	Railway	Wheels	Rear Stack	Side Tanks
5617	1913	1m	BR	0−6−2+0−6−2	Yes	None
5618	,,	,,	,,	,,	,,	,,
5619	,,	,,	,,	,,	,,	,,
5620	,,	,,	,,	,,	,,	,,
5621	,,	,,	,,	,,	,,	,,
5622	,,	,,	,,	,,	,,	,,

The above were designed for the F. C. de Bolivia by the Consulting Engineers Livesey, Son & Henderson and suffered from the problem of having separate coal tenders attached at the smokebox end. When later converted to oil fuel these were to prove exceptionally good locomotives.

Type 3 was a later development from type 1 and in these locomotives the bogies were pushed further apart. Water capacity was increased with either full length side tanks or, in some designs, a divided tank leaving the firebox clear for easy access to the centre of the locomotive. In these instances part of the tank was located behind the cab and the other hung pannier fashion over the front portion of the boiler. All Type 3 locomotives had their cylinders mounted at the outer ends of each steam bogie, this form being introduced in 1908 for locomotives having leading/trailing trucks.

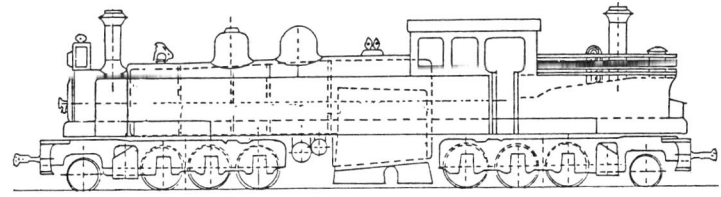

Type 3 was constructed alongside type 1 which continued to be used for 0−6+6−0T until the end of construction of this wheel arrangement in 1921.

Works No.	Year	Gauge	Railway	Wheels	Rear Stack	Tank Style
4580	1908	5ft 5¹³⁄₁₆in	GtS of S	2−8+8−0T	Yes	Full
4581	,,	,,	,,	,,	,,	,,
4582	,,	,,	,,	,,	,,	,,
4735	1910	3ft 6in	ACN&R	2−6+6−2T	,,	,,

Works No.	Year	Gauge	Railway	Wheels	Rear Stack	Tank Style
4736	,,	,,	,,	,,	,,	,,
4841	1911	2ft 6in	A(C)&B	,,	,,	,,
4853	1912	3ft 6in	ACN&R	,,	,,	,,
4854	,,	,,	,,	,,	,,	,,
4972	1913	,,	MR	,,	,,	,,
4973	,,	,,	,,	,,	,,	,,
4974	,,	,,	,,	,,	,,	,,
4975	,,	,,	,,	,,	,,	,,
5400	1927	3ft 0in	CNR	,,	No	Divided
5401	,,	,,	,,	,,	,,	,,
5402	,,	,,	,,	,,	,,	,,
5403	,,	,,	,,	,,	,,	,,
5413	1928	2ft 6in	NWR	2–6–2+2–6–2T	,,	,,
5414	,,	,,	,,	,,	,,	,,
5416	,,	3ft 0in	CNR	2–6+6–2T	,,	Full
5417	,,	,,	,,	,,	,,	,,
5431	1929	,,	,,	,,	,,	Divided
5471	1935	,,	,,	2–8+8–2T	,,	Full
5472	,,	,,	,,	,,	,,	,,

Total construction Type 3 locomotives 23 units.

In the section devoted to Colombian National Railways and in the one covering the Jamaican Government Railways, reference will be found to "pseudo type 3" locomotives. These were not an official Kitson & Co. product but were actually type 1 which had been altered by P. C. Dewhurst to give a cylinder at the outer ends configuration.

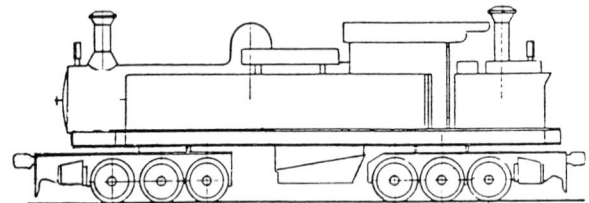

Type 4 represented but two locomotives. Designed by the Consulting Engineers, Livesey Son & Henderson these had three bogies of which only two were driven. The cylinders were placed at the outer ends of each steam bogie.

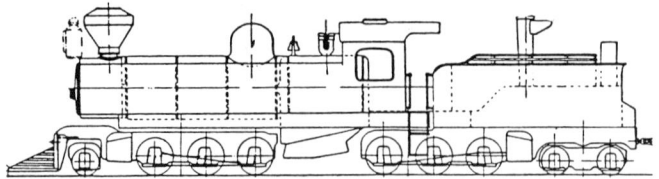

Works No.	Year	Gauge	Railway	Wheels	Rear Stack	Side Tanks
4534	1908	2ft 6in	A(C)&B	2–6+6–4T	No	Full
4568	,,	1m	LR	,,	Yes	,,

Total construction Type 4 locomotives 2 units.
The type was not a success.

Type 5 Kitson Meyer locomotives were a combined rack/adhesion design for use on the Transandine Railway only.

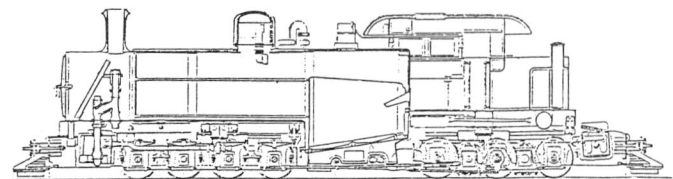

These were built as follows:

Works No.	Year	Gauge	Railway	Wheels	Rear Stack	Side Tanks
4488	1907	1m	CTR	0–8+6–0T	No	Short
4598	1908	1m	,,	,,		
4664	1909	,,	,,	,,	No	Short
4669	,,	,,	ATR	,,	,,	,,
4670	,,	,,	,,	,,	,,	,,
4674	,,	,,	,,	,,	,,	,,
4842	1911	,,	,,	,,		
4882	1912	,,	,,	,,	No	Full
4883	,,	,,	,,	,,	,,	,,

Total construction Type 5 locomotives 9 units.

The principal users of Kitson Meyer articulated locomotives were the South American railroads whose tracks climbed the steep western slope of the Andes, often by near impossible continuous gradients combined with severe curvature. The concept and design was of such excellence as to render the Kitson Meyer the perfect locomotive for mountain operation and indeed operation over lines which ranked as the hardest in the entire world. The South American Kitson Meyer locomotives were subjected to the roughest service expected from motive power anywhere and amazingly one or two survived to the ripe old age of 70+ years, surely a remarkable testimonial to the excellence of the design and construction. The longevity of these machines is even more remarkable when one remembers that Kitson & Co. closed down before the Second World War and for the last 45 years all spares had to be made by the operating company. In South America, Kitson Meyer articulated locomotives were used in Argentina, Brazil, Bolivia, Chile and Colombia. In other parts of the world the type was employed in Jamaica, Spain, Africa, India and Manilla but the type was inextricably bound up with the Andean roads of South America.

By 1910 the Kitson Meyer had developed into a potentially excellent articulated, but despite this the owners of Airedale Foundry were not progressive in their sales policy and the Kitson Meyer was therefore built only in relatively small numbers. Had Kitson & Co. fully developed the Meyer design and been endowed with the marketing policy of say Beyer Peacock & Co. Ltd., then the Kitson Meyer story might well have been totally different.

KITSON MEYER LOCOMOTIVES IN SOUTH AMERICA

These four Chapters relate to the railways in South America which used Meyer/Kitson Meyer type locomotives. The content of each Chapter is as follows:

CHAPTER 2 – CHILE
Anglo-Chilian Nitrate & Railway Co. (FCTT) 3ft 6in gauge
Taltal Railway (FCT)	3ft 6in gauge
Nitrate Railways (FS)	4ft 8½in gauge

CHAPTER 3 – CHILE, BOLIVIA AND BRAZIL
Antofagasta (Chili) & Bolivia Railway (FCAB) 2ft 6in gauge
The Bolivia Railway (FC de B)	1 metre gauge
The Leopoldina Railway (LR)	1 metre gauge

CHAPTER 4 – COLOMBIA
The Colombian Government Railways (FCN):
Girardot Railway	3ft 0in gauge
Cundinamarca Railway	3ft 0in gauge

CHAPTER 5 – ARGENTINE, CHILE
Argentine Transandine Railway (ATR)	1 metre gauge
Chilian Transandine Railway (CTR)	1 metre gauge

CHILE:
ANGLO–CHILIAN NITRATE & RAILWAY CO.
TALTAL RAILWAY
NITRATE RAILWAYS

2

ANGLO CHILIAN NITRATE & RAILWAY COMPANY (FCTT)

The prototype Kitson Meyer locomotive emerged from the Leeds works of Kitson & Company in 1894 to the order of the 3ft 6in gauge Anglo Chilian Nitrate & Railway Company, a South American nitrate carrier serving the districts around Toco. The Company came into being to acquire nitrate grounds located in the Province of Antofagasta and to construct a railway and other works in connection with these acquisitions. The concession granted to this Company by the Chilean Government for construction of the Ferrocarril de Tocopilla al Toco was dated 20 January 1888. Work commenced immediately on the 3ft 6in gauge line connecting the open roadstead Pacific coast port of Puerto de Tocopilla with the Estación Toco, traffic commencing over the whole 87.3km on 15 November 1890. The railway existed simply to convey nitrate from the oficinas on the pampa, down to the port of Tocopilla. The single track line commenced at Tocopilla (52½ft above sea level) and once clear of the yard immediately encountered an immensely difficult climb of the western slope of the Andes involving 17 miles of continuous 1 in 25 grade (actually 1 in 25/24.6/25 and 24.4) of which no less than 75% was combined with fearsome curves of 181ft minimum radius. From Barriles the gradient eased to 1 in 34½ approx for the next 7¼ miles or so to Tigre.

According to *Railways of South America, Part III – Chile*, a branch line was built in 1895 running northwards to Santa Fé. A map reproduced in the above noted volume, and presumed correct (circa 1925), shows the main line from Tocopilla to El Toco, and a north–south line from a crossing place a little west of El Toco, running north to Santa Fé and south to Providencia. The line originally terminated at Estación Toco where the Anglo Chilian Nitrate & Railway Company trans-shipped goods to the 1 metre gauge Longitudinal Railway, but over the years various branch lines were built and when the Maria Elena

nitrate plant was under construction in 1926, the successors to the ACN&R – the Anglo Chilian Consolidated Nitrate Corporation – obtained a Government concession to build a rail line from El Tigre to the Maria Elena plant. A further extension pushed the line to Pedro de Valdivia where another nitrate plant had been established.

For the opening, Kitson & Company (of Leeds) supplied some rigid frame 4–8–4T with 17in × 21in cylinders, 3ft 2½in driving wheels, 160lb boiler pressure and producing 18,900lb tractive effort at 75% boiler pressure. The total weight in working order was 51 tons 5 cwt and with an adhesive weight of 33 tons, Kitson & Company estimated that over over the 1 in 25 grades and 48lb rail of the Ferrocarril de Tocopilla al Toco (FCTT) – to give the railway its then correct title – the 4–8–4T should manage to haul 138 tons of train. During the early 1890s nitrate traffic increased requiring locomotives of greater power and in 1891 the Yorkshire Engine Company (of Sheffield) supplied two Fairlie 0–6+6–0T (Works No's 446/7), these becoming FCTT No's 8 and 9. Little is known regarding the performances of these Fairlie articulated locomotives but both had disappeared from the roster before the Guggenheim take-over in January 1925.

We have noted in the previous Chapter the significant part played by Robert Stirling (Loco Superintendent of the FCTT) in the design and development of the Kitson Meyer articulated, the prototype emerging from the Leeds works of Kitson & Co. in 1894. Two other identical locomotives followed in the same year and one more in 1895, all being shipped to Chile where three of the four (one was lost at sea) were put to work on the hill between Tocopilla and Barriles. The new locomotives proved extremely successful hauling 115–125 ton trains, the difficult 17 mile climb taking 2½ hours including two water stops. One problem seems to have been the considerable

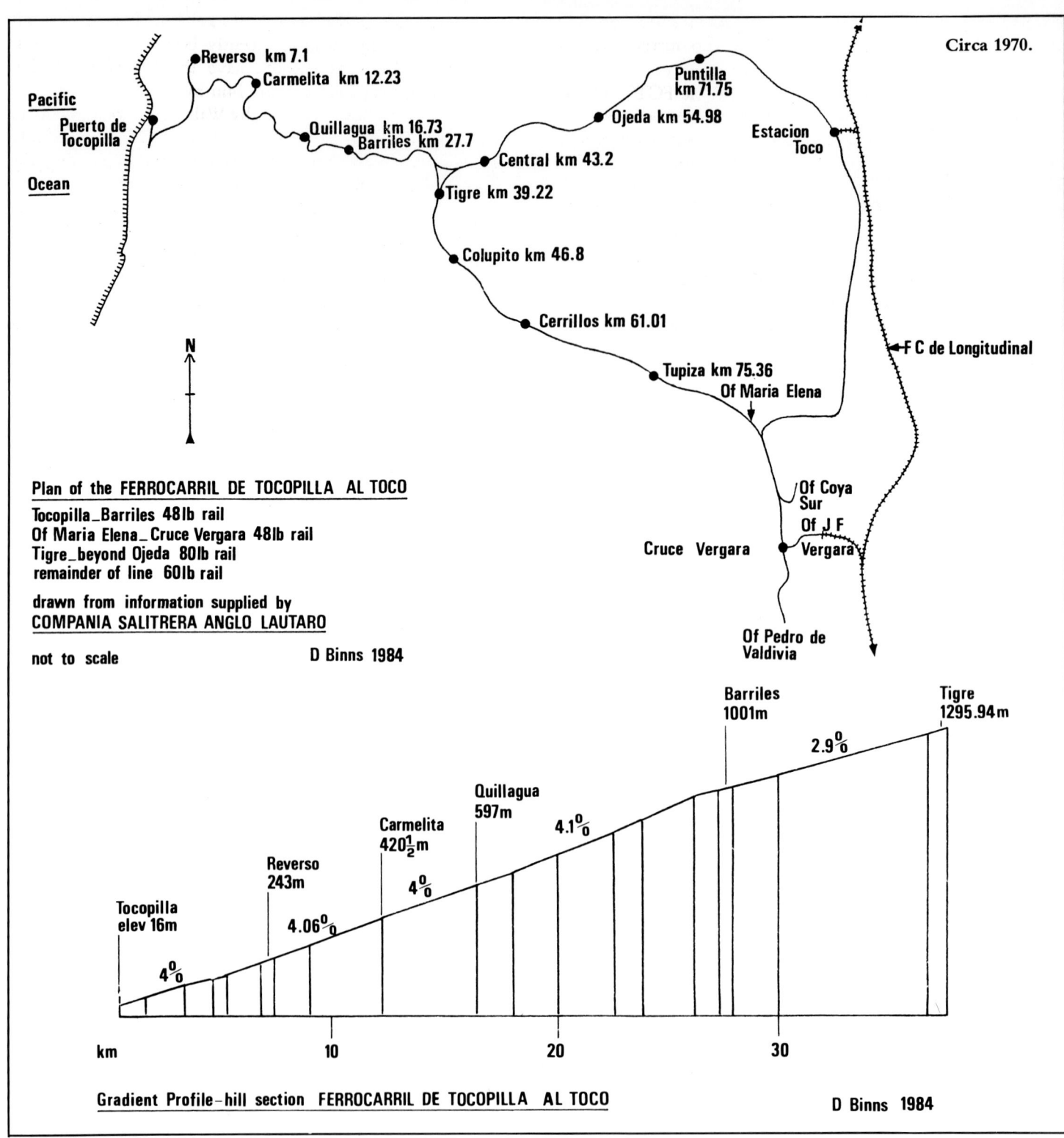

Plan of the FERROCARRIL DE TOCOPILLA AL TOCO

Tocopilla_Barriles 48lb rail
Of Maria Elena_Cruce Vergara 48lb rail
Tigre_beyond Ojeda 80lb rail
remainder of line 60lb rail

drawn from information supplied by
COMPANIA SALITRERA ANGLO LAUTARO

not to scale   D Binns 1984

Gradient Profile-hill section FERROCARRIL DE TOCOPILLA AL TOCO   D Binns 1984

overhang of the side tanks on these narrow gauge articulateds which made stability on curves something of a hazard according to Andean locomotive inspectors who had experience with the type. According to Brian Fawcett in *Railways of the Andes*, on one occasion a Kitson Meyer had rolled over on a sharp curve on the steep seaboard grade and gone down the mountainside in the days before electrification. Another problem was that the makers claim that the boilers were easily washed out was not corroborated by men on that job.

When Kitson & Co. shipped these locomotives the purchasing company was given a set of working drawings and a strange turn of events occured on 11 November 1902 when Kerr Stuart & Co., California Works, Stoke-on-Trent, was given an order for "one double bogie Meyer locomotive with 4 cylinders 14in × 18in in accordance with specifications and drawings". Why Kerr Stuart & Co. was given this order is not known—it may have been due to Kitson's embarrassment at the near copies they had produced of the Baldwin locomotive. It is not known if Kitson & Co. and the FCTT had had dialogue on this subject resulting in the latter approaching a different locomotive builder, but on the other hand possibly the FCTT wanted this locomotive quickly and Kitson & Co. were simply unable to oblige due to pressure of work. Alternatively it could well have been a matter of price since it is recorded that Kerr Stuart & Co. lost about £2000 on this locomotive. In the event they produced no further Kitson Meyer type locomotives. The Kerr Stuart Meyer was built from Kitson & Co. drawings supplied by the FCTT and as may well be expected was virtually the same as the four previous units. It was shipped in 1903. At this point reference must be made to an article in the *Locomotive Magazine* for 1903 in which a second Meyer articulated was noted as being built by Kerr Stuart & Co. Works No. 817 is sometimes erroneously quoted as being the second Meyer but was actually "one 14in six coupled double ender (2−6−2T) shunting engine in accordance with specifications and drawings". This was for the FCTT but was not articulated.

A further four Kitson Meyer locomotives were built at Leeds in 1909 for the FCTT and like all their predecessors were 0−6+6−0T type 1 simple locomotives with the cylinders at the rear of each steam bogie. All had rear chimneys when built. *The Locomotive* for 15 March 1910 had this to say:

"ANGLO-CHILIAN NITRATE RYS.−Four 0−6−6−0 articulated tank engines have recently been built by Messrs. Kitson & Co. Ltd., which have the following leading dimensions:−cylinders (four) 14in by 18in, with flat-sided valves placed above, actuated by Walschaerts gear; diameter of coupled wheels 2ft 10¾in; length of main frame 33ft 2in; boiler: length 10ft 8⅞in, diameter 4ft 4in, containing 206 tubes of 1¾in diameter; working pressure 160lb per sq in; total capacity of side and bunker tanks 2,040 gallons of water; coal space 110 cubic ft, oil fuel 80 cubic feet. In working order the engines weigh 62 tons 14 cwt, and they are fitted with a combination vacuum brake, sight-feed lubricators, cow catchers and central buffer."

These locomotives were equipped to burn coal or oil as required.

By this time the FCTT had 8 type 1 0−6+6−0Ts, but 2 additional Kitson Meyers built in 1910 and 2 more in 1912 were modified type 3 2−6+6−2T with larger cylinders, increased

For the opening of the Anglo Chilian Nitrate & Railway Company's line Kitson & Co. supplied four rigid frame 4−8−4T which were able to haul 138 tons of train over the 1 in 25 grades. Two more of the same type were supplied in 1902. R. N. Redman collection.

In 1891 the ACN&R Co. purchased two 0−6+6−0T Fairlie articulated locomotives but all subsequent articulated power was of the Kitson Meyer type. ACN&R Co.

This Kitson Meyer 0–6+6–0T was built by Kitson & Co. for the ACN&R Co. in either 1894 or 1895 and may well have been one of the first three. The photograph shows to advantage the early Type 1 locomotives with cylinders at the rear of each bogie. Note the firebox dropped down between the bogies. A comparison with the photograph reproduced on page 3 reveals minor differences—the locomotive on this page has a full diameter rear chimney (as opposed to the small diameter one shown on the locomotive on page 3), the bell has not been fitted yet, but the leading footsteps have. Of interest is the special open cab designed to give the engine crew some relief from the terrific heat of the desert.

R. N. Redman collection.

FCTT

Works No's 3532/3/4 of 1909

Tare 40.90

Loaded 55.35 Adhesive

| 8.55 | 8.90 | 9.65 | tons | 8.80 | 9.75 | 9.70 |

heating surface and grate area. With 32,810lb of tractive effort at 75% boiler pressure, these were substantial and powerful narrow gauge locomotives. *The Locomotive* for 15 November 1910 reported thus:

"ANGLO-CHILIAN NITRATE RYS.—This company has lately added two large articulated tank engines to its rolling stock. The wheel arrangement is 2−6−6−2, the coupled wheels are 3ft 2½in in diameter, while the leading and trailing wheels which are carried in radial boxes, are 2ft 0½in in diameter. The cylinders are four in number, and are 15in diameter by 21in stroke; the valves are of the flat slide pattern and are driven by Walschaerts gear. The boiler barrel is 5ft 2½in in diameter and 11ft 6in long, containing 281 tubes of 1¾in outside diameter; working pressure 180lb per sq in. The heating surface is made up as follows: firebox 196.21 sq ft, tubes 1,528.64 sq ft, total 1,654.85 sq ft; grate area 34 sq ft; wheel base of engine 38ft 9in; from rail to top of chimney is 12ft 10in. The engines are built to burn coal or oil, and have a capacity of 130 cubic ft for coal and 95 cubic ft for oil; the tanks carry 2,500 gallons of water. In working order they weigh 77 tons 17 cwt. Fittings include screw reverse, sanding, O.P. whistle, tank water gauge, central buffer and the vacuum brake."

These 2−6+6−2T were a development from type 1. In type 3 the bogies were pushed further apart and leading and trailing trucks were incorporated into the design. The cylinder position was altered to the outer ends of each bogie, this form being introduced by Kitson Co. in 1908 on 3 large 2−8+8−0T supplied to Spain. These 2−6+6−2T were originally used on the hill section but by 1939 had been relegated to work on the pampa between Maria Elena and Pedro de Valdivia oficinas. Their tonnage rating was 480.

Kitson Meyer locomotives supplied to the AC&NR

0−6+6−0T Kitson & Co. Works No's 3532/3/4 of 1894. Steam trial dates 28/6, 27/7 and 30/8/1894. Shown as Meyer class in Kitson records. Type 1 simple locomotives with cylinders at the rear of each steam bogie.

0−6+6−0T Kitson & Co. Works No. 3604 of 1895. Steam trial 22/7/1895. Type 1 simple locomotive with cylinders at the rear of each steam bogie.

0−6+6−0T Kerr Stuart & Co. Works No. 816 of 1903. Type 1 simple locomotive with cylinders at the rear of each steam bogie. Named "Toco Pilla" when new.

0−6+6−0T Kitson & Co. Works No's 4653/4/5/6 of 1909. Type 1 simple locomotives with cylinders at the rear of each steam bogie.

2−6+6−2T Kitson & Co. Works No's 4735/6 of 1910. Type 3 simple locomotives with cylinders at the outer ends of each steam bogie.

2−6+6−2T Kitson & Co. Works No's 4853/4 of 1912. Steam trials 20/12/1911 and 9/1/1912. Type 3 simple locomotives with cylinders at the outer ends of each steam bogie.

Total 13

AC&NR Locomotive list

This list is a joint venture produced by myself, Mel Turner and Reimar Holzinger and lists all the known locomotives purchased between 1889 and 1911.

The solitary Kerr Stuart built Kitson Meyer type 0−6+6−0T for the ACN&R Co. This was built from drawings supplied to the Company by Kitson & Co.
D. Binns collection.

Road No.	Wheels	Maker	Works No.	Year Built
1	4–8–4T	Kitson & Co.	3185	1889
2	4–8–4T	Kitson & Co.	3186	1889
3	4–8–4T	Kitson & Co.	3187	1889
4	4–8–4T	Kitson & Co.	3188	1890
5	0–6–4T	Manning Wardle	1139	1889
6	0–4–0ST	Manning Wardle	1107	1889
7	0–4–0ST	Manning Wardle	1126	1889
8	0–6+6–0T	Y. E. Co. Fairlie	446	1891
9	0–6+6–0T	Y. E. Co. Fairlie	447	1891
10	0–6+6–0T	Kitson Meyer	3532	1894
11	0–6+6–0T	Kitson Meyer	3533	1894★ lost at sea?
11 second	0–6+6–0T	Kitson Meyer	3534	1894 ex-No. 12
12	0–6+6–0T	Kitson Meyer	3534	1894 renum. 11
12 second	0–6+6–0T	Kitson Meyer	3604	1895
13	2–6–2T	Kitson & Co.	3535	1894
14	2–6–2T	Kitson & Co.	3536	1894★ lost at sea?

★Works No's 3533/6 are noted in a list prepared by Mike Page together with the note "11 and 14 vanish (lost at sea?) and 12 becomes 11". This could well be the case—probably both were shipped on the same vessel.

Road No.	Wheels	Maker	Works No.	Year Built
14 second	2–6–2T	Kitson & Co.	3601	1895
15	2–6–2T	Kitson & Co.	3613	1895
16	0–4–2T	Kitson & Co.	3977	1900
17	0–4–2T	Kitson & Co.	3978	1900
18	4–8–4T	Kitson & Co.	4108	1902
19	4–8–4T	Kitson & Co.	4109	1902
20	2–6–2T	Kerr Stuart	817	1903
21	0–6+6–0T	Kitson Meyer built by Kerr Stuart	816	1903
22	2–6–2T	Kitson & Co.	4340	1905
23	0–6+6–0T	Kitson Meyer	4653	1909
24	0–6+6–0T	Kitson Meyer	4654	1909
25	0–6+6–0T	Kitson Meyer	4655	1909
26	0–6+6–0T	Kitson Meyer	4656	1909
27	0–4–2T	Avonside E. Co.	1581	1910
28	0–4–2T	Avonside E. Co.	1582	1910
29	2–6+6–2T	Kitson Meyer	4735	1910
30	2–6+6–2T	Kitson Meyer	4736	1910
31	2–6–2T	Kitson & Co	4839	1911
32	2–6–2T	Kitson & Co.	4840	1911
33	2–6–2T	Kitson & Co.	4857	1911
34	2–6–2T	Kitson & Co.	4858	1911
35	2–6–2T	Kitson & Co.	4859	1911
36	2–6+6–2T	Kitson Meyer	4853	1911
37	2–6+6–2T	Kitson Meyer	4854	1911

The Company was operated as originally organised until 22 December 1924, when the Anglo-Chilean Consolidated Nitrate Corporation was incorporated under the laws of the State of Delaware to acquire all the property of the Anglo-Chilian Nitrate & Railway Co. (Ltd.). The new company was registered in Chile as the Compania Salitrera Anglo-Chilena. The principal parties interested in the new concern were the Guggenheim Bros., of 120 Broadway, New York City.

Rolling stock purchased by the Guggenheim Bros. consisted of 35 steam locomotives and about 600 wagons.

Locomotives

	Wheels	Year	Road No's
6	4–8–4T	1889/90, 1902	No's 1–4, 18, 19
1	0–6–4T	1889	No. 5
2	0–4–0ST	1889	No's 6–7
8	0–6+6–0T	1894/5, 1903/09	No. 10, 11, 12, 21, 23–26
10	2–6–2T	1895, 1903/05/11	No. 13–15, 20, 22, 31–35
4	0–4–2T	1900/10	No. 16, 17, 27, 28
4	2–6+6–2T	1910/12	No. 29, 30, 36, 37
——			
35			
——			

The 2 Fairlie 0–6+6–0T purchased in 1891 were not taken over by Guggenheim and must have been disposed of previous to the take-over in January 1925. These carried Road No's 8 and 9.

By 1928 there was a surplus of steam locomotives due to the electrification of the steeply graded coastal escarpment and from that date a slow run-down was effected bringing steam operation to an end in 1959. One rather strange acquisition, at some unknown date between 1927 and 1938, was an 0–6–6–0 Mallet tender locomotive which carried No. 105. This is thought to have been built for Russia as one of a batch of 50 not delivered due to the Revolution. The subsequent history of these locomotives is clouded with complications and will not be persued here. Possibly this Mallet was previously owned by a Guggenheim company and was simply transferred to the FCTT. At any event, it was not logical to increase the steam locomotive fleet at a time when the ACN&R Co. had a surplus of steam anyway.

The following is extracted from the *Railway Magazine* for September 1928:

"THE BRAKE QUESTION IN CHILE

The Anglo-Chilean Consolidated Nitrate Corporation's railway, which comprises 100 miles of 3ft 6in gauge main line, serves the port of Tocopilla. The principal business is to haul nitrate from the different 'oficinas' to the port of Tocopilla. The line has about 28km of 40 per cent grade which the fully-loaded convoys must descend. The rolling-stock at the time of purchase consisted of 35 steam locomotives and about 600 wagons.

Because of the adverse operating conditions, the type of automatic brake originally fitted could not be depended upon to control the trains down the grade, and the hand brakes had been used exclusively for the grade operation, necessitating the employment of one man to every three cars in a convoy. When the Guggenheim interests took over the operation of this railway in 1925, their engineers made a complete study of the railway, which included a thorough investigation of the braking question, and after numerous and exhaustive trials, decided to adopt the Westinghouse automatic brake.

Last November, operation of the trains fitted with Westinghouse air brakes was begun. It was proved to be very successful. The system installed is known as the Westinghouse empty and load brake, automatic with non-automatic control. The brake was designed to develop a total brake shoe pressure equal to 75 per cent of the empty weight of the car, when the wagons are empty or partly loaded, and 40 per cent of the total weight when the wagon is fully loaded. These ratios are based on 50lb brake cylinder pressure. Nearly all the railways in the Andes having heavy

One of the 1910 pair (Works No's 4735/6), Type 3 2−6+6−2T for the ACN&R Co. Note that the bogies are pushed further apart with the cylinders at the outer ends of each unit. These locomotives were equipped to burn either coal or oil.   *D. Binns collection.*

No. 29 from the 1910 batch of Type 3 2−6+6−2T was delivered in the same condition as the locomotive pictured above, but when photographed its rear tank side sheets had been raised. The cab had been enclosed in this circa 1939 photograph and No. 29 had received American fittings including Westinghouse compressor, air reservoir and brake equipment, generator, electric headlamp and central coupling. The front sandbox had been moved forward to allow space for the air compressor. This photograph was taken at Maria Elena depot.   *B. Fawcett.*

grades are now using Westinghouse automatic brakes with non-automatic control."

An undated specification sheet provided by the Sociedad Quimica y Minera Chile, SA, states that 7 0−6+6−0T supplied by Kitson & Co. were in service, but since the document carries the initials CSAC for Compãnia Salitrera Anglo Chilena, the date could not have been before 1931. The 7 then in service carried Road No's 10, 11, 12, 21, 24, 25 and

26. No. 23 may have been sold to the Taltal Railway as a replacement for one of its 0−6+6−0T. Steam was withdrawn from the FCTT in 1959 following which some of the serviceable 0−6+6−0Ts and 2−6−2T were disposed of to the Taltal Railway—for details please see section on the Taltal Railway.

Unfortunately disposal details relating to the 4 2−6+6−2T are unknown.

Railway	FCTT	FCTT	FCTT	FCTT	FCTT	FCTT
Gauge	3ft 6in	3ft 6in	3ft 6in	3ft 6in	3ft 6in	3ft 6in
Wheels	0–6+6–0T	0–6+6–0T	0–6+6–0T	0–6+6–0T	2–6+6–2T	2–6+6–2T
Maker	Kitson	Kitson	Kerr Stuart	Kitson	Kitson	Kitson
Works No.	3532/3/4	3604	816	4653–6	4735/6	4853/4
Year	1894	1895	1903	1909	1910	1912
Cylinder position	rear of each bogie	rear	rear	rear	outer ends	outer ends
Cylinders–inches	14 × 18	14 × 18	14 × 18	14 × 18	15 × 21	15 × 21
Boiler Pressure–lbs/sq in	160		160	160	180	180
Heating Surface:						
Firebox–sq ft	104	104	98	104	126.2	126.2
Tubes–sq ft	1064	1064	1044	1042	1528.6	1528.6
Total–sq ft	1168	1168	1142	1146	1654.8	1654.8
Superheater–sq ft	None	None	None	None	None	
Total HS–sq ft	1168	1168	1142	1146	1654.8	
Grate area–sq ft	25.25	25.25	24.5	25.5	34	34
Driving wheel–dia	2ft 10¾in	2ft 10¾in	2ft 10¾in	2ft 10¾in	3ft 2½in	3ft 2½in
Other wheels–dia	None	None	None	None	2ft 0½in	2ft 0½in
Rigid wheelbase–first group	6ft 2½in	6ft 2½in	6ft 2½in	6ft 2½in	7ft 0in	7ft 0in
Rigid wheelbase–second group	6ft 2½in	6ft 2½in	6ft 2½in	6ft 2½in	7ft 0in	7ft 0in
Total wheelbase	25ft 6½in	25ft 6½in	25ft 6½in	25ft 6½in	38ft 9in	38ft 9in
Water–galls	1900	1900	1860	2040	2500	2500
Fuel tons	3	3	3	80cu ft oil 110cu ft coal	95cu ft oil 130cu ft coal	
Weight empty–tons/cwt	40–18		46–5	48–8		
Weight in W.O.–tons/cwt	55–7	55–7		62–13	77–17	79–18
Weight Adhesive–tons/cwt	55–7	55–7		62–13	61–10	62–11
Tractive effort–lbs	27,600 (85%)		27,586 (85%)			32,810 (75%)
Overall–height	12ft 10in	12ft 10in	12ft 10in	12ft 10in	12ft 10in	
” –width	8ft 4in	8ft 4in	8ft 4in		8ft 10in	
” –length	36ft 9½in	36ft 9½in		36ft 11in		
Hauling capacity on straight level track at 8–10mph–tons				2707		
On 1 in 100 @ 8–10mph				705		
On 1 in 75 @ 8–10mph				556		
On 1 in 50 @ 8–10mph				353		
On 1 in 25 @ 8–10mph				179		

No. 10 was the prototype Kitson Meyer (Works No. 3532/1894). It was photographed in 1939 in Tocopilla yard in up-graded post-Guggenheim condition with Alliance 6in × 6in draft gear, steam and Westinghouse brakes, Westinghouse air compressor, generator and electric lighting.   B. Fawcett.

The Kitson Meyer locomotives detailed above are confirmed but there are a number of highly improbables relative to the FCTT. Lionel Wiener in his book *Articulated Locomotives* mentions one other 0−6+6−0T from the works of Kerr Stuart & Co. in 1909. The dimensions quoted do not agree with those for Kerr Stuart Works No. 816 of 1903 but despite extensive research all attempts to trace this locomotive have failed. Wiener also mentions some 2−6+6−2T type 1 simple locomotives with the cylinders at the rear of each steam bogie supplied by Kitson & Co. in 1903/4. These are most unlikely since the use of leading and trailing pony trucks would make it impossible for the steam bogie units to be identical with this cylinder layout. Wiener quotes dimensions basically as the later 1910/12 units but differing in firebox heating surface and a slight variation in the working order weight. Attempts to trace these supposed locomotives have not unexpectedly failed and this entry by Wiener is almost certainly a statistical inexactitude and should not be given credence.

There is also considerable doubt concerning a possible 2−6+6−2T with cylinders at the outer ends of each steam bogie−the type 3 modification−reputedly supplied by Kitson & Co. in 1905. This was mentioned by Brian Fawcett in the book *Railways of the Andes*: "A further batch was supplied by Kitson in 1905, this time with rather more power. These had outward facing cylinders and pony trucks on each unit. When I was on the FCCT in 1939, this latter class was at work on the pampa between Tigre and Maria Elena. The tonnage rating of the Kitson's was 480". In a letter, Brian Fawcett told me that he thought there were 3 of the original 0−6+6−0T (confirmed) and when he was on the FCTT there were as far as he could recall 5 2−6+6−2T working on the pampa, from Maria Elena sheds (4 confirmed). The later ones had the same dimensions and the valves, valve gear, rods etc. of these were interchangeable with a class of 2−6−2T switchers (No's 13, 14, 15, 22, 31−35) by Kitson & Co., at work in the Maria Elena yards. They (the 2−6+6−2T) were much heavier engines with 180lb boiler pressure and greater tractive effort. It seems highly unlikely that Kitson & Co. produced a type 3 locomotive as early as 1905 and indeed if such a locomotive did exist it would of course have been the first of its type. Against this possibility is the fact that Kitson continued supplying 0−6+6−0T type 1 locomotives to the FCTT until 1909. Nor does this supposed locomotive appear in Kitson's surviving records, nor can the Sociedad Quimica y Minera de Chile confirm its existence. All in all it seems highly unlikely that a type 3 locomotive was built as early as 1905 and that Fawcett has quoted a wrong date.

From research it is felt that all the above, improbables did not exist.

CHILE−THE TALTAL RAILWAY (Ferrocarril Taltal)

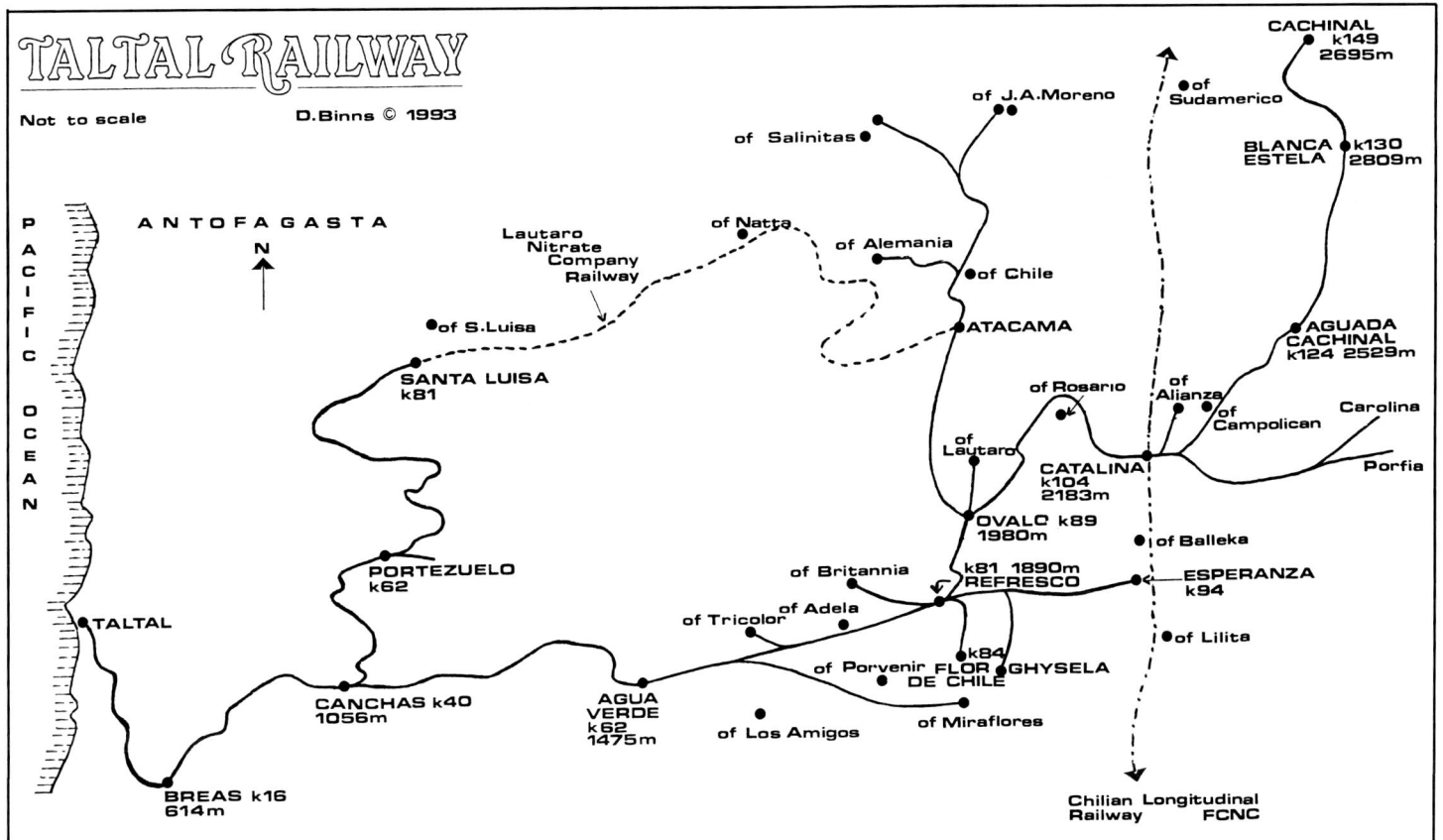

The Taltal had the distinction of being the last railway in the world to operate Kitson Meyer articulated locomotives. Located in the so-called "rainless belt", this 174 mile 3ft 6in gauge railway served the southern part of the Province of Antofagasta, in northern Chile, the principle reason for its existence being nitrate.

The history of the Taltal Railway (Ferrocarril Taltal) dates back to 1872 but it was not until 14 November 1878 that Concession No. 284 was granted. Supplementary decrees of 17 February and 12 March 1880 amended the original concession to some extent and the modified version was transferred by the original holder Alfredo Quaet Falsem, to the Taltal Railway

Ferrocarril Taltal 0−6+6−0T No. 50 was Kitson & Co. Works No. 4288/1904 and the first Kitson Meyer delivered to this railway. Although generally similar to the Anglo Chilian Nitrate & Railway Kitson Meyers, the Taltal units had more conventional cabs and were typical of the early Type 1 locomotives.
D. Binns collection.

Company, incorporated in London on 3 June 1881. The new railway was constructed by Henry Meiggs—a US railway builder of some note—and the line opened from the Pacific coast port of Taltal, to Refresco (81km) in October 1882. An additional concession dated 17 October 1887 authorised an extension of the main line to Cachinal—opened June 1889. Other branch lines followed:

Santa Luisa branch authorised by concession dated 18 January 1890

Atacama branch authorised by concession dated 2 February 1893

Valliera branch authorised by concession dated 21 October 1894

Miraflores branch authorised by concession dated 21 October 1894

Chile and Alemania branch authorised by concession dated 31 July 1903

Morena branch authorised by concession dated 22 December 1905

Additional concessions were authorised for the construction of other smaller branches and at 30 June 1928 the Company had 280km of track in operation.

The line commenced at the Pacific coast port of Taltal (39ft above sea level) climbing to 9217ft at Blanca Estela 83.3 miles distant, by way of 1 in 20/21 grades and 426ft radius curves passing through spectacular barren scenery.

During the mid-1920s both coal and oil fuel were in use, there being one coal station holding 240 tons and three oil stations with a total capacity of 45 tons—these being located about 23¼km apart. Water was supplied by the railway company to the town of Taltal, as well as providing water for the ships in the bay and the railway's own needs—much of this being piped in from further up the line. Along the railway there were 8 water stations spaced some 35¾km apart.

The Taltal Ralway purchased its locomotives between the years 1881 and 1907—these comprised 8 0−6−0T, 15 2−6−0T, 6 2−6−2T, 2 0−6−4T and 10 Kitson Meyer 0−6+6−0T supplied between 1904 and 1907.

Kitson Meyer locomotives supplied to the F. C. Taltal

0−6+6−0T Kitson & Co. Works No. 4288 of 1904. Steam trials 15/10/04. Road No. 50.

0−6+6−0T Kitson & Co. Works No's 4432/3/4 of 1906. Steam trials 19/12/06, 9/1 and 1/2/1907. Road No's 51, 52 and 53.

0−6+6−0T Kitson & Co. Works No's 4504/5/6 of 1907. Steam trials 27/9, 31/10 and 11/10/1907. Road No's 54, 55 and 56.

0−6+6−0T Kitson & Co. Works No's 4512/3/4 of 1907. Steam trials 16/11, 21/11 and 6/12/1907. Road No's 57, 58 and 59.

All ten were simple tank locomotives of type 1 with the cylinders at the rear of each steam bogie and all had rear chimneys.

The Kitson Meyer's were employed on nitrate trains from the interior to the port of Taltal and at the peak of the system no less than 15 trains departed from Taltal for the various "oficinas" each morning, returning loaded later the same day. In addition, a passenger service was operated between Taltal and the metre gauge Longitudinal Railway at Catalina and the various "oficinas" were also served by passenger trains.

The Lautaro Nitrate Company built its own railway from Santa Luisa to Atacama where connection was made at both ends with the Taltal Railway. The Lautaro line was operated in connection with the Taltal Railway so that a complete circuitous route was formed from Canchas to the nitrate fields.

Following the cessation of steam on the FCTT in 1959, a number of its serviceable locomotives were sold to the Taltal Railway—presumably the Kitson Meyer's included in the sale would have been dismantled into three main units for shipping by sea, and then re-assembled at Taltal.

Besides 5 known 2−6−2T, Tocopilla disposed of probably 6 0−6+6−0T. From the time of arrival on the Taltal Railway it becomes almost impossible to identify specific locomotives by works numbers. No official details of locomotives transferred or received seems to exist, nor is it possible to ascertain anything from the works plates carried by Taltal 0−6+6−0T since these

Kitson & Co. Works No. 4504/1907 was Taltal Railway No. 54. The date of this photograph is unknown but was taken during the years of prosperity. The rear bunker has been raised on this Type 1 Kitson Meyer.

J. Agnew collection.

were changed frequently with wrong plates being used all the time. All we can do is to present the information available.

0−6+6−0T supplied to Tocopilla−new:

Maker	Work's No.	Year Built	Road No.
K	3532	1894	10
K	3534	1894	11
K	3604	1895	12
KS	816	1903	21
K	4653	1905	23
K	4654	1905	24
K	4655	1905	25
K	4656	1909	26

0−6+6−0T supplied to Taltal−new:

Maker	Work's No.	Year Built	Road No.
K	4288	1904	50
K	4432	1905	51
K	4433	1906	52
K	4434	1906	53
K	4504	1907	54
K	4505	1907	55
K	4506	1907	56
K	4512	1907	57
K	4513	1907	58
K	4514	1907	59

Official records show that on 28 December 1970, ten Kitson Meyer 0−6+6−0T remained on the Taltal Railway, numbered as follows: 50, 54, 55, 57, 57, 58, 59, 59, 60, 61. These ten were shown in official records but one additional locomotive was seen that year—No. 51. What follows is a combination of official information and sightings made by visitors to the railway—it does not offer any concrete results and probably we will never know which locomotive was which.

Taltal Railway Kitson Meyer Locomotives at 28/12/1970

No's

50 Seen in steam earlier in 1970. Originally thought that this was Works No. 4288—the original Kitson Meyer supplied to the Taltal Railway. From photographic evidence, I now think this locomotive was ex-FCTT 1909 build (with increased water capacity). FCTT No. 23 is a possibility—Kitson 4653. Seen on a 1977 visit in the shed—bunker only.

54 Not seen during 1970 visit. Probably Works No. 4504. In fact not seen on any subsequent visit.

55 Seen during 1970 visit. Probably Works No. 4505. In 1977 dumped minus boiler.

57 Probably Works No. 4512. Not seen on 1970 visit. Seen in shed in 1977 carrying Kitson 1907 plates—almost certainly Works No. 4512.

57 Ex-FCTT (No. 12 or 21). Only one No. 57 was seen on the occasion of the 1970 visit—this locomotive was inscribed CSF. CHL (probably the initials of the Flor de Chile Nitrate Company—a concern served by the F. C. Taltal. Why this locomotive carried these initials is not known—maybe it had been sold to them? In 1977 this locomotive is recorded as dumped.

17

A Taltal Kitson Meyer shunts wagons of sacked nitrate into a storage shed at the Pacific Coast port of Taltal.
D. Binns collection.

No. 50 was moving No. 61 from the workshop to the running shed when photographed in 1970. Differences are obvious between the two locomotives: No. 61 has a straight topped tank whereas No. 50's tank is of increased capacity. Bunkers are also different with No. 61 having high solid sheets. No. 50 may well have been 1909 built ex-FCTT No. 23 whilst No. 61 is believed to be an earlier locomotive.   J. Wiseman.

No. 50 at Taltal in 1970. No. 61 on the extreme left. Note the generator and electric head and back-up lights. The rear stack on No. 50 has been cut down.   J. Wiseman.

58   Not seen during 1970 visit. Probably Works No. 4513. Seen dumped, tanks only in 1977.

59   Not seen during 1970 visit. Ex-FCTT No. 26 Works No. 4656. Photographic evidence shows this locomotive to have extended tank to give increased water capacity. See

Note ★

59   Ex-No. 56. Possibly Works No. 4506 but why renumbered to 59 remains a mystery. Not seen during 1970 visit. See Note ★

★ One of this pair was serviceable in 1977, the other was

on the dump.

60   Ex-FCTT No. 10. Wrecked and minus one bogie. Kitson 1907 plates. If this locomotive was in fact FCTT No. 10, it was the prototype Kitson Meyer supplied in 1894 and should have had 1894 plates. There is however, doubt as to the origins of No. 60 since both No. 10 and No. 60 were seen in 1977. In 1977 No. 60 was seen on shed—its boiler had exploded in 1969.

61   Ex-FCTT No. 25 Works No. 4655. Under repair in 1970. Carried both Kitson 1907 and 1909 plates, so obviously little of value can be ascertained from any works plates fitted at this period. Not seen in 1977.
I don't think this is ex-FCTT No. 25—suspect it to be an earlier locomotive. In service 1972. Not seen 1977.

In addition to the 10 0−6+6−0T on official records, one additional locomotive was seen in 1970:

51   Ex-FCTT. Presumably a replacement for the original No. 51—rusty condition. Possibly Works No. 4654. In the works 1977.

On the occasion of this visit, five further locomotives were said to be locked in a separate shed and some of the unseen 0−6+6−0T may have been amongst them. Obviously by 1970 all the surviving locomotives must have been a mixture of several others as no spare parts had been available since before the Second World War. In 1972 No. 61 was in service and reported as carrying Kitson 1906 works plates and still retaining its rear chimney had changed little in 65 years. The 1906 works plate highlights the unreliability when it comes to identifying locomotives since No. 61 had both 1907 and 1909 plates when seen two years earlier. In view of the discrepancies which occur with regard to works plates, it is felt that little credence should be given to these and in all probability in later years any works plate was attached without regard to accuracy. Another thought is that the No. 61 seen in 1970 was not the same No. 61 seen two years later.

In 1977 the following additional Kitson Meyer locomotives were observed:

10   Ex-FCTT. In the works.

51   Taltal Railway original. Works No. 4432. In the works.

?   Also the boiler and frame of one other 0−6+6−0T.

By this time identification of surviving locomotives was impossible with two No. 51's, two No. 57's and two No. 59's to confuse the issue. Both No. 51's, were together in the works in the course of overhaul, one clearly numbered 51 on its side tanks, the other revealing the same number following cleaning of its bunker. Further complications were brought about because in the years 1930 on, locomotives were often loaned or borrowed from the FCTT and these were renumbered on transfer. Unfortunately no details survive as to these transfers.

By 1976 the Taltal Railway was all but closed down and a visit made in the following year revealed that the end was indeed at hand and that the last train "to the interior" had run the previous October.

In 1979 a communication from the Chilian Ministry of Minerals & Chemicals intimated that the Taltal operation had closed down and the railway was to be dismantled. On a visit made in 1979 only one Kitson Meyer was still serviceable—No.

59, and this in deplorable condition. Also in evidence were the remains of many Kitson Meyer's with pieces of plate frames, cylinders, wheels, side tanks, piles of boiler tubes and several rear bunkers with chimneys still attached. Two other Kitson Meyer locomotives were in evidence standing out in the open amongst piles of corrugated iron sheeting and wood roof rafters from the dismantled engine shed—one being jacked up for repairs which were never finished, its side tanks bearing the inscription CSF. CHL, so this may well have been No. 57 which was reported during the 1970 visit. The other Kitson Meyer seen in 1979 was unidentified (probably No. 60) but its boiler had exploded in 1969 and no repairs had been carried out.

Although its origin are uncertain (probably ex-FCTT No. 26—Works No. 4656), Kitson Meyer 0−6+6−0T No. 59 of the Taltal Railway was unquestionably the last of these articulated locomotives to run anywhere in the world and its final job seems to have been operation of the demolition train used to dismantle the line. The wrecking crew worked from east to west and when all the track was lifted No. 59 was isolated in the yard on the coast at Taltal. In 1979 No. 59 was scheduled as a historic monument, but little appears to have been done to save the locomotive. It was in steam for a short run in March 1986. A letter dated 27 March 1986 from E. F. Clark comments "re- the preserved Taltal loco—this is in rather bad state having been severely vandalised and caught up in local politics". On 30 November 1989 John C. Verser wrote "The Taltal Kitson Meyer is sinking in the sand where it is on display, almost down to the running gear".

In August 1990 Ian Thomson (Santiago) wrote "Meanwhile, I decided last weekend that it was time to start a scandal here in Chile, to get something done about the Taltal Kitson-Meyer. I sat down Sunday and wrote an article, in Spanish, and threatened to have it published in a local news-type magazine. This seems to be having the desired reaction, for Eulogio Gordo, the guy who has come to own the locomotive, according to an intermediate source, is prepared to sell it. And the mayoress of Taltal has told me she is prepared to let it go to the Santiago Railway Museum.

I hope that there'll be some good news regarding the locomotive within a few weeks".

At the time of writing (February 1993) it is still missing from the Museum, apparently uncared for and seemingly unwanted. Obviously the Santiago Museum is where No. 59 should be but one cannot help feeling that this Kitson Meyer would have fared better had it been brought back to this country to rest either in the National Railway Museum or in the Armley Mills Museum at Leeds, a fitting home for an interesting locomotive type developed and built in Leeds.

The recent history of Taltal Railway No. 59, in brief
Ian Thomson, 10 February 1993

1.   The Taltal Railway ceased commercial operations in 1976, when the Alemania Nitrate Factory closed.

2.   The line was taken up and all metal objects, including locomotives, were cut up for scrap between 1978 and 1980. No. 59 was saved by being declared a National Monument through Decree No. 1221 of the Ministry for Public Education. I have no idea who asked for it to be declared a National Monument. It was neither me nor the Asociacion Chilena de Conservacion del Patrimonio

FC Taltal No. 61 was photographed in green livery with white lining, red pilot beams, balance weights, motion and red smokebox door hinge and details, heading a train of nitrate towards Taltal in brilliant desert evening sunlight in 1972. R. Christian.

Ferroviario, which did not exist at that time. I presume it must have been the then Mayor of Taltal.

3. No. 59 was used on the track taking-up train and was steamed once or twice for the benefit of photographers.

4. I saw the locomotive in the Taltal depot in January 1978, when it was still being used on dismantling operations, and in January 1980, when it was still operable and accompanied by some nitrate-carrying wagons and most of the old station.

5. I then presumed it was being tidied up for display in Taltal's main square. My folks were vacationing in Chile in Summer 1986. With my father I drove to Arica, and came back slowly, to see things on the way. One of the things I wanted to show him was No. 59 proudly displayed in the square. It was not there. Some locals told us it was still where it had been since 1980, i.e. in the ex-station, which by then had been dismantled all around it. It was in a shabby state and had already lost most bronze

In 1972 No. 61 was in service. When photographed in 1970 it was devoid of lining but here sports a complete lining job. The origin of this locomotive is uncertain—it was not one of the original Taltal 0−6+6−0T and is probably ex-FCTT. R. Christian.

and copper parts. The locals told me the company which owned the railway, and which had taken it up, had gone bankrupt and its property was being put up for auction by receivers.

6. On getting back to Santiago, I went to see the lawyers concerned, who had already received one bid, for $400,000, from a scrap metal merchant. I told them it was a National Monument, which forbids sale for scrap. (One of the problems with National Monuments in Chile is that nobody knows what has been declared a National Monument.) They took note of the matter, and the next I heard was that they were planning to auction it at much higher ($700,000) minimum price, banking on the fact that its historical significance would increase its sale value.

7. At the auction, no bids were offered. The lawyers lowered the price to $500,000 and tried again. This time, the ACCPF decided to formerly intervene, preferably by getting the locomotive preserved in a dignified way in the North. First of all, I called up possibly interested companies in the Second (Antofagasta) Region, to see whether they would want to buy it. I tried (amongst others) the FCAB and Soquimich, the owner of the Tocopilla to Toco railway and the only Nitrate Factories remaining in operation, but nobody was the slightest bit interested. Thus we decided to have a try at preserving it in the Quinta Normal Museum. We had very little money so, at the suggestion of the Association's Secretary at that time, Alberto Araneda—I was President—we had printed 900 shares and tried to sell them at $1,000 each. We got very good press and TV coverage but only managed to sell 350 of them.

8. Even so, we would have made it, since no other bids were in sight. On the morning of the second auction, set for 12.00 hrs. the Mayor of Taltal called up a contractor who had done quite a bit of construction work in the zone (Manuel Gordo, of the company Eulogio Gordo y Cia.) and asked him to buy it, adding that the Municipality of Taltal would buy it from him once it had raked together enough money to do so. Gordo made the only bid, and bought it for $500,000.

9. The campaign launched by the ACCPF attracted the eyes of the Chilean Post Office, which got in touch with the Asociación and proposed our collaborating on issuing a special stamp with the locomotive as its theme. This was duly done, priced at the then airmail letter tag of $95. The stamp was launched in Antofagasta in December 1986, the Regional Governor (Intendente) and myself doing the honours.

10. Gordo did nothing at all with the locomotive, which continued to rot away. By March 1987 it had lost all bronze and copper parts, its front headlamp had gone, somebody had pierced its fuel tank to drain away whatever fuel oil was left and even some of the valve gear parts had been stolen.

11. By 1990, the Mayor had been switched for a Mayoress (María Teresa Cornide), a nice lady who invited me to tea one Sunday morning in May 1990, at her wooden house by the seafront. She said she couldn't invest public money in restoring the locomotive, since it was private property. She wrote to Gordo about the locomotive, but had received no reply by September.

12. I managed, finally, to speak by phone with Manuel Gordo, in February 1991. He said he still expected the Municipality to buy the locomotive from him, but added that he would be prepared to sell it to the ACCPF as long as the Mayoress gave her approval. (The Asociación would have the locomotive shipped to a safer place, out of Taltal such as the FCAB Museum in Antofagasta, and Gordo was worried that he might irk the local population if he did something which resulted in the locomotive being shifted away from Taltal against its wishes.)

13. She never approved the sale to the ACCPF, and the locomotive is stuck in a legal impasse.

Railway	Taltal	Taltal
Gauge	3ft 6in	3ft 6in
Wheels	0−6+6−0T	0−6+6−0T
Maker	Kitson	Kitson
Works No.	4288	*4432/3/4 1906 †4504/5/6 1907
Year	1904	†4512/3/4 1907
Cylinder Position	rear of each bogie	rear of each bogie
Cylinders−inches	14 × 18	14 × 18
Boiler Pressure−lbs/sq in	160	160
Heating Surface:		
Firebox−sq ft	107	107.25
Tubes−sq ft	1075	1042.40
Total−sq ft	1182	1149.65
Superheat−sq ft	None	None
Total−sq ft	1182	1149.65
Grate area−sq ft	25.2	25.2
Driving wheel diameter	2ft 10¾in	2ft 10¾in
Other wheels	None	None
Rigid wheelbase−first group	6ft 2½in	6ft 2½in
Rigid wheelbase−second group	6ft 2½in	6ft 2½in
Total wheelbase	25ft 6½in	25ft 6½in
Water−galls	1865	1865
Coal−tons	3	3
Weight empty−tons/cwt		46 4
Weight in W.O.−tons/cwt	60 17	60 17
Weight Adhesive−tons/cwt	60 17	60 17
Tractive effort−lbs		
Height−overall		12ft 10in
Width − "		8ft 6¼in
Length− "		35ft 5in
Hauling capacity on straight level track at 8−10mph	*2920 tons	†2964 tons
On 1 in 100 @ 8−10mph	*723 tons	†734 tons
On 1 in 75 @ 8−10mph	*500 tons	
On 1 in 50 @ 8−10mph	*391 tons	†397 tons
On 1 in 25 @ 8−10mph	*183 tons	†186 tons

Taltal 0−6+6−0T No. 59 at Taltal in 1977 with the Pacific Ocean behind. The rocky terrain gives the town of Taltal its name—literally 'the place of many stones'.
                                                                                                              A. E. Durrant.

No. 59 at work in the Atacama desert, where it reputedly never rains although some very sparse vegitation is evident. Note the black smoke from the front chimney and white steam from the rear cylinders exhausted through the separate rear chimney.
                                                                                                              A. E. Durrant.

No. 59 at Taltal.   A. E. Durrant.

Opposite side of Taltal No. 59 standing in the old passenger station before setting off for the run through the Atacama desert.   A. E. Durrant.

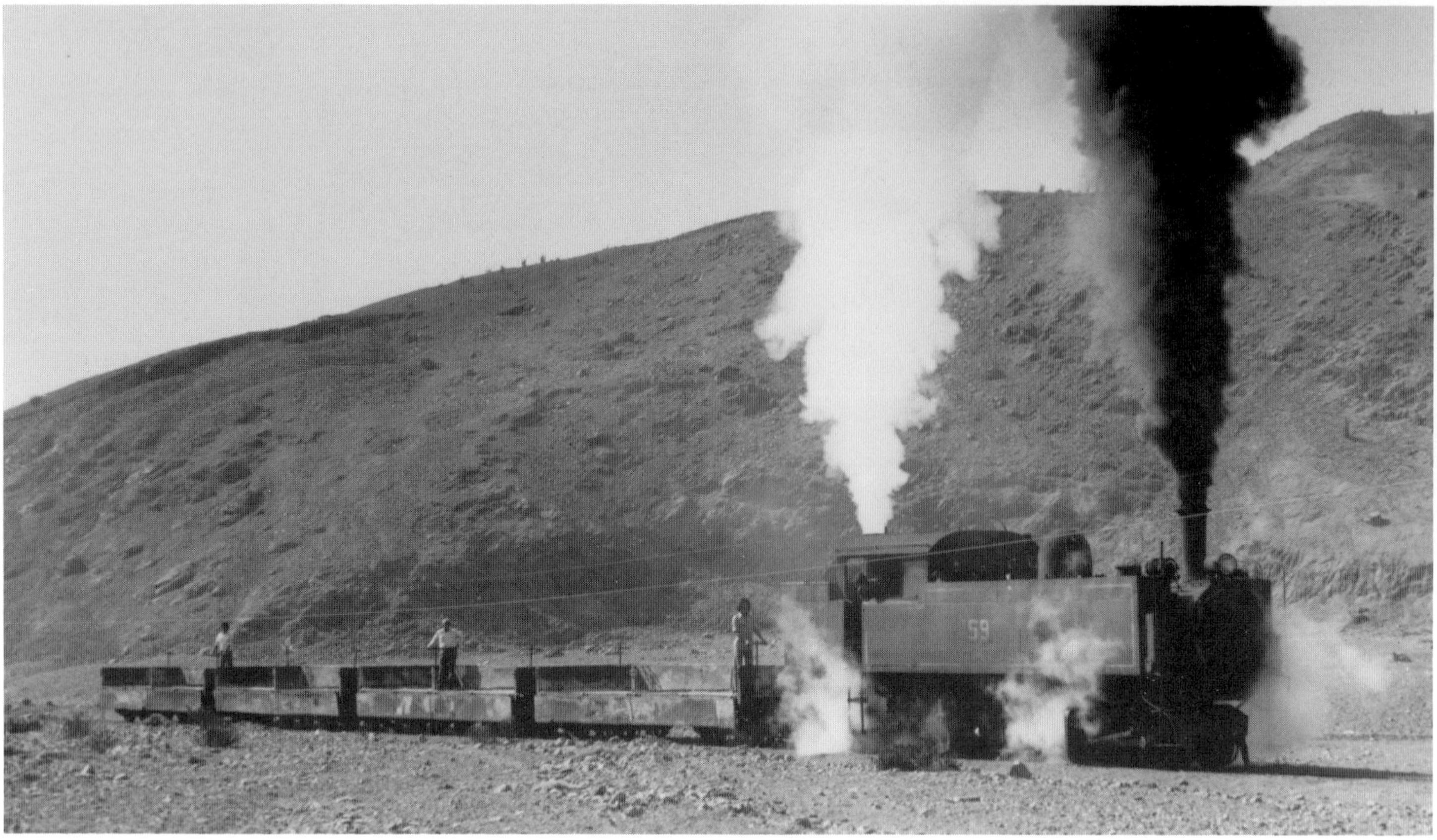
59

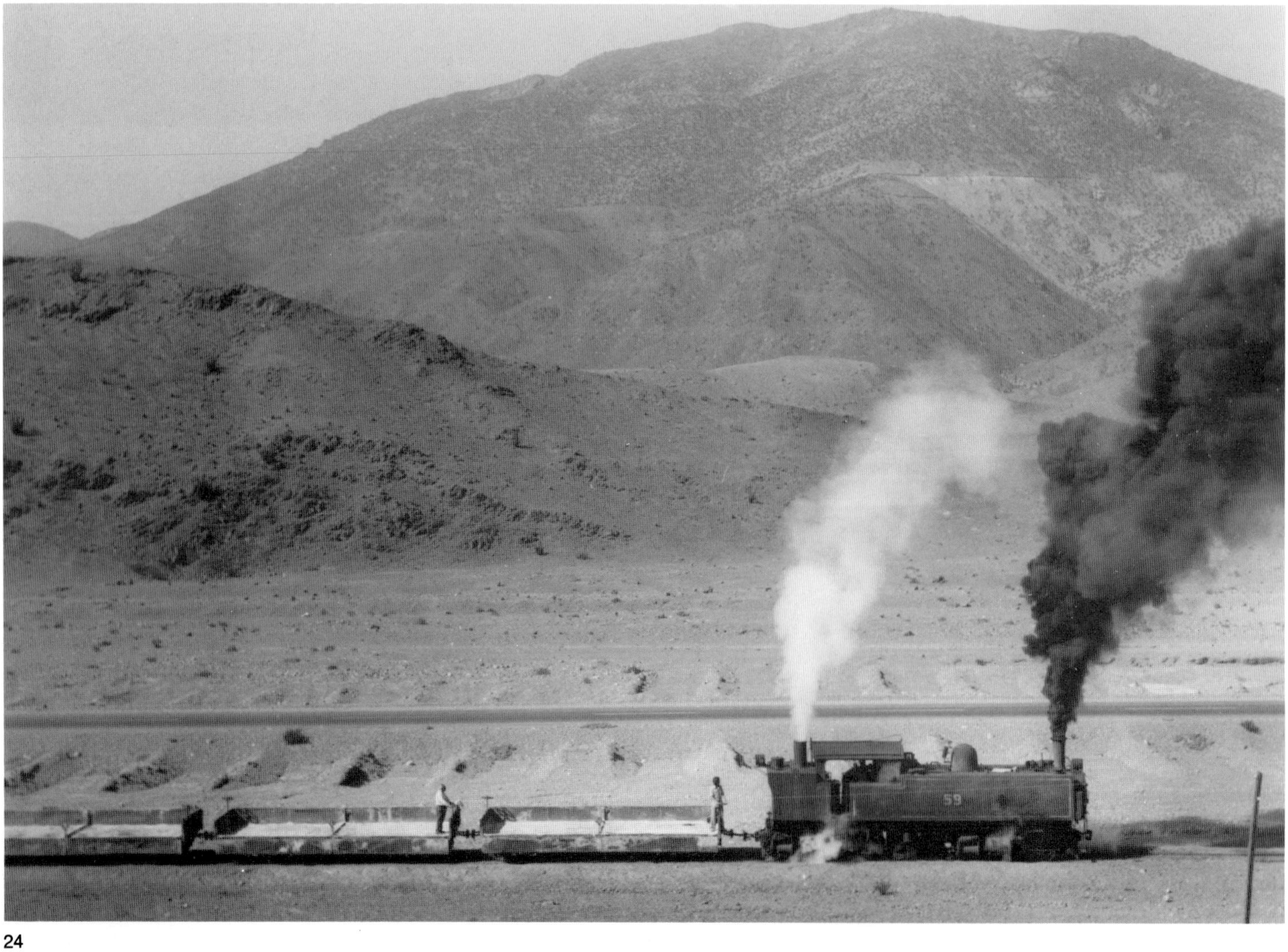
59

Rear view of Kitson Meyer 0–6+6–0T Taltal Railway No. 59 on its train at Taltal station in 1977.   A. E. Durrant.

Opposite page: Two views of Kitson Meyer 0–6+6–0T No. 59 in the Atacama desert 1977. These were staged run-pasts with 'fumo negro' from the first chimney and saturated steam from the rear, a Kitson Meyer speciality. On the 1 in 24 grade with no continuous brakes, each wagon's brakeman is at the ready.   A. E. Durrant.

No. 59 in the abandoned station at Taltal in 1980.   R. Thomson via I. Thomson.

Head on view of Taltal No. 59 showing bogie frame offset from boiler on curve. Also clear are the welded enlargements to the original rivetted water tanks.
A. E. Durrant.

No. 59 out of use at Taltal January 1990.
I. Thomson.

No. 59 at Taltal in May 1990.   *I. Thomson.*

CHILE—NITRATE RAILWAYS LTD.
(El Ferrocarril Salitero)

On 11 June 1860 the Peruvian Government granted a concession to Federico Pezet and Jose M. Costa for the construction and operation of a railway between Iquique and La Noria. This concession lapsed and in 1864 a similar one was granted to Jose Pickering and Manuel Orihuela. No work was achieved and the concession passed into the hands of the Ramon Montero Bros. Company, who built the 4ft 8½in gauge Iquique—La Noria line, a distance of 73 miles between 1868 and 1871. It was known as the Pisagua—Lagunas Railway & Branches and opened for traffic on 27 July 1871. Early motive power included a Fairlie 0−6+6−0T named "Tarapaca", acquired from the Fairlie Engine & Steam Carriage Company, to haul 150 ton trains over eleven miles of 1 in 26 grade. A second Fairlie 0−6+6−0T "Hercules" was purchased from the Avonside Engine Company in 1871.

On 18 May 1869 the Ramon Montero Bros. Co. received an additional concession for the construction of a 90 mile line from Pisagua to Zapiga. This was also operated by Fairlie locomotives, three being supplied by the Avonside Engine Co. in 1871/2. A further concession authorised the construction of "certain branch railway lines" as well as a highway to connect with the Bolivian border. The war of the Pacific occurred and the lines owned by the Montero Bros. passed to Chile.

On 24 August 1882 the Nitrate Railways (Ltd.) was registered in London and with the consent of the Chilean Government took over the holdings of the Ramon Montero Bros. Co. Further extensions were built including 95 miles to the Lagunas nitrate area—this opening in January 1893. The Nitrate

Railways invested heavily in Fairlie articulated locomotives, purchasing both 0−6+6−0T and 5 2−6+6−2T, these latter eventually being converted back to 0−6+6−0T. The Company also worked Porter 2−8−2T and some massive Yorkshire Engine Company 4−8−4T which could haul 160−180 ton trains at 7½mph.

During 1908 the Company took delivery of two extremely large and powerful Meyer type simple articulated locomotives from the Yorkshire Engine Company, Sheffield, and these were used along with the Fairlie's over the difficult 19¼ mile section between the Pacific coast port of Iquique and Las Carpas. From Iquique (26ft above sea level) El Ferrocarril Salitero ran almost level for 2½ miles before starting a near impossible 19¼ mile climb of the Andes to a point 3,000ft above sea level at Las Carpas. Very heavy grades existed with almost negligible tangents separating the 173 curves which occur on this section. The sharpest curve was 280ft radius and at one point 3 curves of 300, 350 and 800ft radius follow one upon the other—these being on a grade of approx. 1 in 21. The principal grade was 1 in 25 compensated and at 22km a short section of 1 in 23.3 existed. The entire 19¼ miles was a continuous climb and the economical operation of trains was always a problem. There was heavy downhill traffic in nitrate and during the mid-1920s on the Iquique—Las Carpas section eight locomotives each made three round trips daily handling an uphill tonnage of 3,000 gross tons per day. In order to keep traffic moving several trains were run close together in the same direction and for a period of about 90 minutes trains

Nitrate Railways No. 101 (YE Co. Works No. 1945/1923); one of six massive 4–8–4T built by the Yorkshire Engine Company (Works No's 1941-6 of 1923/4) and numbered 97-102 in the NR fleet. These were built to do the same work as the earlier Fairlie articulated 0–6+6–0T and the two Meyer 0–6+6–0T and were followed in 1926 by three 2–8–2+2–8–2T Beyer Garratt's which at that time were the most powerful in the world.
D. Binns collection.

Yorkshire Engine Company Works No. 941/1908 was one of a pair of large Kitson Meyer type 0–6+6–0T which became Nitrate Railways No's 74 and 75. The year of construction was 1908 and not 1912 as noted in my previous book.
D. Binns collection.

departed from Iquique at periodic intervals and five or more could be seen climbing the mountain at the same time. Most of the Fairlie articulateds were used on this section along with both of the Meyer 0−6+6−0T.

0−6+6−0T Yorkshire Engine Co. Works No's 940/1 of 1908.

Meyer type simple locomotives with the cylinders at the rear of each steam bogie. NR Road No's 74 and 75. Scrapped about 1930.

The Meyer's were required to haul 208 ton trains at an average speed of 8½mph without stopping for fuel or water. At the time of their construction these were very large and powerful locomotives with four 17in × 22in cylinders, 3ft 9in driving wheels and 180lb boiler pressure producing 43,322lbs of tractive effort. A rear chimney exhausted steam from the rear unit and a 27 element feed water heater was located in the lower part of the rear water tank.

The Nitrate Railways purchased no further Meyer or Kitson Meyer type locomotives, later turning to Beyer Garratt articulateds, three 2−8−2+2−8−2 being ordered in 1926 and at that time with a tractive effort at 85% boiler pressure of 78,370lbs, were the most powerful in the World. A further three were supplied in 1928.

The Nitrate Railways had invested heavily in Fairlie articulated locomotives (probably 34 were purchased) and all but five were supplied as 0−6+6−0T. The 2−6+6−2T illustrated had been built as one of ten 0−6+6−0T, five of which were destined for the Poti & Tiflis Railway (Russia). The other five, Yorkshire Engine Company Works No's 224-228/1874, were for a time unsold, being eventually converted to 2−6+6−2T for service on the Nitrate Railway−to which they were shipped in 1882. All five were subsequently converted back to 0−6+6−0T. D. Binns collection.

Railway	Nitrate Railways of Chili
Gauge	4ft 8½in
Wheels	0−6+6−0T
Maker	Yorkshire Engine Co.
Works No.	940/1
Year	1908
Cylinder Position	rear of each bogie
Cylinders−inches	17 × 22
Boiler Pressure−lbs/sq in	180
Heating Surface:	
Firebox−sq ft	171
Tubes−sq ft	2135
Total−sq ft	2306
Superheater−sq ft	None
Total HS−sq ft	2306
Grate area−sq ft	39.3
Driving wheel−diameter	3ft 9in
Other wheels−diameter	None
Rigid wheelbase−first group	8ft 6in
Rigid wheelbase−second group	8ft 6in
Total wheelbase	35ft 8in
Water−galls	4000
Coal−tons	4
Weight empty−tons/cwt	90 0
Weight in W.O.−tons/cwt	117 2
Weight Adhesive−tons/cwt	117 2
Tractive effort−	43,322
Height−overall	13ft 6in
Width − ”	10ft 0in
Length− ”	49ft 9in

CHILE, BOLIVIA AND BRAZIL:
ANTOFAGASTA (CHILI) & BOLIVIA RAILWAY
THE BOLIVIA RAILWAY
THE LEOPOLDINA RAILWAY

3

ANTOFAGASTA (CHILI) & BOLIVIA RAILWAY (FCAB)

Historically the 2ft 6in gauge Antofagasta (Chili) & Bolivia Railway dates back to 1872 when a concession was granted by the Government of Bolivia (at that time the Province of Antofagasta formed part of the Republic of Bolivia) to Melbourne Clarke & Company for the construction of a 2ft 6in gauge railway from Antofagasta to Melbourne Clarke's nitrate grounds at Pampa Alta. They then founded the Compañia de Salitres y Ferrocarril de Antofagasta (Antofagasta Nitrate & Railway Company) in Valparaiso, Chile, on 9 October 1872 and transferred the railway concession to that company. Rail was first laid in 1873 and the first section of 18.6 miles (29.9km) to Salar del Carmen opened in 1st December that year. Initially, mules were used as motive power to bring the caliche (nitrate bearing soil) down to the Pacific coast. The railway was extended to Salinas in 1879—the start of the Antofagasta nitrate pampa at 5,000ft above sea level. It was further extended to Central in 1882, and to Pampa Alto (150km) in 1883.

The Compañia Huanchaca de Bolivia, owners of the Huanchaca Silver Mines, eventually became owners of the railway and in 1888 the undertaking was sold to the British owned Antofagasta (Chili) & Bolivia Railway Company, but was not worked by them until 1904. The Company owned considerable mileage and its main line ran from Antofagasta (previously part of Bolivia but claimed by Chile in the 1879 war) and the nearby port of Mejillones, to Oruro (in Bolivia), a distance of 925.273km. From Oruro, metre gauge leased track took the FCAB to Viacha and forward to La Paz. By the mid-1920s the Company owned 813 miles of track and leased a further 417 miles from the Bolivian Railway Company as well as operating over 441 miles of the Chilian Longitudinal Railway. Two separate adminstrations were responsible for the line—one based in Antofagasta covered the railways in Chile, whilst the other at La Paz was responsible for the Bolivian section.

Built to the narrow gauge of 2ft 6in the FCAB was no miniature railway, carrying in 1925 over 1½ million tons of freight on the Chilian section and more than ¼ million tons on the Bolivian side. Principal freight down to the Pacific coast ports was nitrate, bar copper, borate and other ores with oil, fuel, coal, coke, machinery, timber and general merchandise forming the bulk of the inland traffic. On the Bolivian section downhill traffic included tin barilla, lead and silver slags and silver ore with general merchandise, fuel and mining machinery in the reverse direction. There were few important engineering works on the FCAB with no tunnels and only one bridge of any size. The main line from Antofagasta climbed gradually from near sea level to 3,955 metres at Ascotan which was the highest station on its main line. From this point the

track dropped slightly to Ollague—3,696 metres, 436.57km from Antofagasta and the International Frontier crossing point into Bolivia.

From Ollague the main line continued on high table land undulating at elevations varying between 3,806 and 3,658 metres to Oruro. On the Chilian section the minimum grade was 1 in 33 and 393ft radius curves. The route was single with passing loops and no signals—movement of trains being controlled by an electric staff system. From Antofagasta to Oruro the track was originally 2ft 6in gauge and from Oruro to Viacha metre gauge leased track carried trains forward to La Paz. Naturally this change of gauge was an operating problem and delays occurred when freight had to be trans-shipped between the two, this being achieved by transferring the 2ft 6in car bodies onto 1 metre gauge trucks and vice-versa. Eventually it was decided to convert the entire railway from 2ft 6in to 1 metre and the first section to be altered was that between Oruro and Uyuni in 1916. By 1928 all the remaining 2ft 6in gauge main lines had been widened.

Kitson Meyer locomotives supplied to the FCAB
2−6+6−4T Kitson & Co. Works No. 4534 of 1908. Steam trials 28/4/1908. Type 4 simple locomotive with cylinders at the outer ends of each bogie. Livesey Son & Henderson type. Carrying FCAB Road No. 36 it was named "Hercules", being renumbered 38 between 1908 and 1922. Scrapped in 1929 at the time of change of gauge in Bolivia, there being considerable difficulties in altering this locomotive.
2−6+6−2T Kitson & Co. Works No. 4841 of 1911 Steam trials 29/9/1911. Type 3 simple locomotive with cylinders at the outer ends of each bogie. Road No. 37.

The Antofagasta (Chili) & Bolivia Railway Company was not a big user of articulated locomotives. Their first incursion into Kitson Meyer articulation was an order received by Kitson & Co. on 7 June 1907 for one 2−6+6−4T, type 4 simple locomotive with its cylinders at the outer ends of each steam bogie and which had been designed by the Consulting Engineers Livesey Son & Henderson. The locomotive had a leading pony truck, two steam bogies and one non-powered 4-wheel truck beneath the rear of the tender. At the smokebox end the superstructure was mounted on a pivoted frame as in a conventional Kitson Meyer, but the cab end carrying the bunker was also pivoted and rested on the front end of the tender frame beneath which was the rear steam bogie guided by a 4-wheel non-power truck. Equipment included a full length side tank, Westinghouse quick acting brake, bell, siren and whistle, central buffing gear, tank water gauge, screw reversing gear, sanding gear, sight feed oilers and a Belpaire firebox. No rear chimney was provided. This locomotive was shipped on 14 May 1908 entering service later that year but was reported as spending more time off the road than on.

In 1911 a conventional Kitson Meyer 2−6+6−2T was purchased and carrying FCAB No. 37 entered service in the following year. This too had but a short life being scrapped in 1929 at time of gauge conversion. A full length water tank was provided as was a rear stack. In Kitson's records this locomotive was shown simply as an "articulated 2−6+6−2T".

Unfortunately we do not know if these two locomotives were used on the Bolivian or Chilian sections but it is believed they were purchased for use in Chile. Apart from the main route there were several branch lines laid with 50lb rail and with severe gradients and curves of 400ft radius—the Chuquicamata branch (1 in 21.6), the Abra branch (1 in 28) and the Callahuasi branch (1 in 32.4), this latter reaching 15,833ft above sea level. The Chuquicamata and Abra branches were very short and it is unlikely the Kitson Meyer's were used there, the Callahuasi branch being a distinct possibility with a climb of 1,124 metres from the main line at Ollague, to Punto Alto.

The London office of the Antofagasta (Chili) & Bolvia Railway Company Ltd. was extremely helpful to my enquiries but unfortunately our correspondence was dated 1971—just one year too late since the voluminous official records relating to operations in Bolivia were disposed of in London in 1970 and with it were full details of these Kitson Meyer locomotives, information now unfortunately lost forever. One unsolved problem concerning FCAB Kitson Meyer locomotives relates to a statement of Lionel Wiener in his book *Articulated Locomotives*: "this railway which already owned some Kitson Meyer 6 coupled total adhesion locomotives put into service a more . . ." Total adhesion indicate 0−6+6−0T type and as far as can be traced no such ever operated on the FCAB. Certainly no record of these appear in Kitson's list and the Company was unable to find any trace. No further Kitson Meyer locomotives were purchased but the FCAB did purchase three 4−8−2+ 2−8−4 Beyer Garratt locomotives in 1929 and a further six in 1950. An approach to the Company archivist in Antofagasta did not produce any reply.

Railway	FCAB	FCAB
Gauge	2ft 6in	2ft 6in
Wheels	2−6+6−4T	2−6+6−2T
Maker	Kitson	Kitson
Works No.	4534	4841
Year	1908	1911
Cylinder Position	outer ends	outer ends
Cylinders—inches	14 × 18	14 × 18
Boiler Pressure—lbs/sq in	180	180
Heating Surface:		
Firebox—sq ft	102.2	129
Tubes—sq ft	1054.3	1663
Total—sq ft	1156.5	1792
Superheat—sq ft	None	None
Total HS—sq ft	1156.5	1792
Grate area—sq ft	25.1	30
Driving wheel—diameter	3ft 1½in	3ft 1½in
Other wheels—diameter	2ft 3in	2ft 3in
Rigid wheelbase	7ft 0in	7ft 0in
Wheelbase—first group	12ft 9in	12ft 9in
Wheelbase—second group	14ft 7½in	12ft 9in
Total wheelbase	40ft 11½in	40ft 6in
Water—galls side tanks	1225	3100
rear	2525	
Coal—tons	4½	4
Weight empty—tons/cwt	60 15	66 0
Weight in W.O.—tons/cwt	85 0	89 11
Weight Adhesive—tons/cwt	59 13	70 10
Tractive effort—lbs	25,380 (75%)	25,475 (75%)
Height—overall	11ft 6in	12ft 3in
Width − "	8ft 0in	8ft 3½in
Length− "	49ft 7¾in	50ft 6½in
Tender:		
Water—galls.	None	None
Coal—tons	None	None
Wheel diameter	None	None
Total length engine and tender	−	−
Hauling capacity on straight		
level track @ 8−10mph	2838	3061
On 1 in 100 @ 8−10mph	726	741
On 1 in 75 @ 8−10mph	568	576
On 1 in 50 @ 8−10mph	386	389
On 1 in 25 @ 8−10mph	171	169

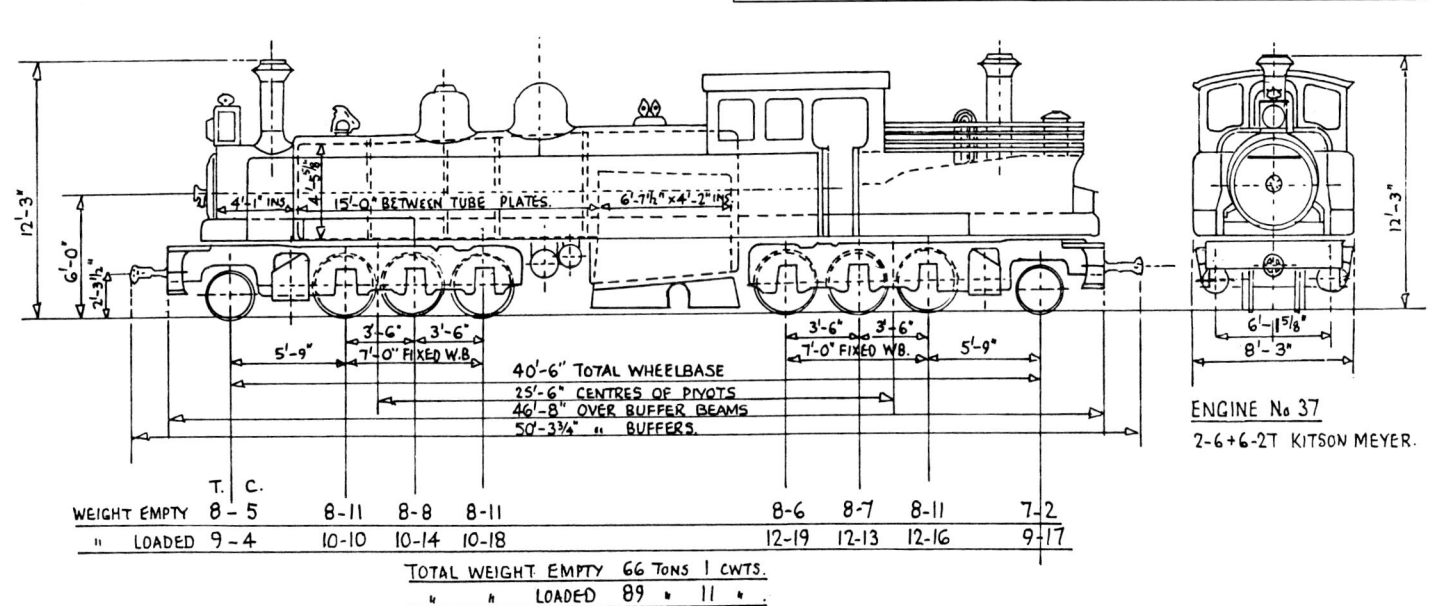

ENGINE No 37
2−6+6−2T KITSON MEYER.

Livesey Son & Henderson Meyer 2−6+6−4T built by Kitson & Company, Works No. 4534/1908 for the FCAB. This locomotive was of the rare Type 4 configuration, a type which extended to two locomotives only − the second being used on the Leopoldina Railway, Brazil. The FCAB example was later renumbered 38 and spent more time off the track than on.
D. Binns collection.

FCAB conventional Type 3 Kitson Meyer 2−6+6−2T, Works No. 4841/1911.
D. Binns collection.

Six 0−6−2+0−6−2 Meyer locomotives were built by Beyer Peacock & Co. Ltd. (Works No's 5617-22/1913. They were Type 2 simple with cylinders at the rear of each bogie. As built these were coal burners but all were subsequently converted to oil.
A(C)&BR Co.

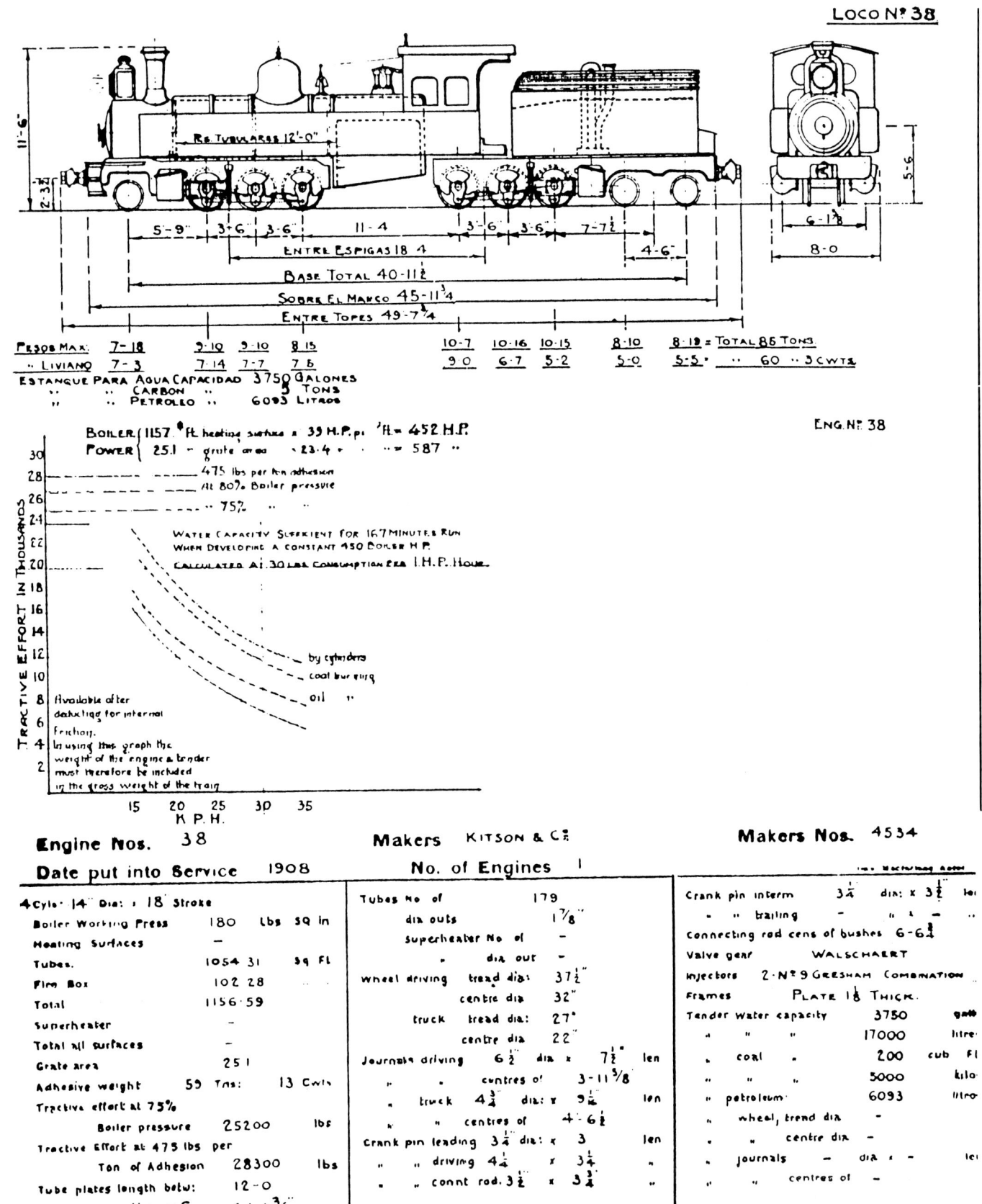

ENG. Nº 38

Engine Nos.	38	Makers	KITSON & Cº	Makers Nos.	4534

Date put into Service	1908		No. of Engines	I	

4 Cyls· 14" Dia: × 18 Stroke

Boiler Working Press	180	lbs sq in	Tubes No of	179		Crank pin interm	3¼	dia: × 3¼	len
Heating Surfaces	—		dia outs	1⅞"		" " trailing	—	" × —	"
Tubes.	1054·31	sq ft	superheater No of	—		Connecting rod cens of bushes	6-6¼		
Fire Box	102·28		" dia out	—		Valve gear	WALSCHAERT		
Total	1156·59		wheel driving tread dia:	37½"		Injectors	2·Nº 9 GRESHAM COMBINATION		
Superheater			centre dia	32"		Frames	PLATE 1⅛ THICK		
Total all surfaces	—		truck tread dia:	27"		Tender water capacity	3750	gall	
Grate area	25·1		centre dia	22"		" " "	17000	litre	
Adhesive weight	59 Tns:	13 Cwts	Journals driving 6½ dia × 7½" len			" coal "	200	cub ft	
Tractive effort at 75%			" " centres of	3-11⅜		" " "	5000	kilo	
Boiler pressure	25200	lbs	" truck 4¾ dia × 9½ len			" petroleum	6093	litre	
Tractive effort at 475 lbs per			" " centres of	4-6½		" wheel, tread dia	—		
Ton of Adhesion	28300	lbs	Crank pin leading 3¼ dia × 3 len			" " centre dia	—		
Tube plates length betw:	12-0		" " driving 4¼ × 3¼ "			" journals — dia × —			len
dia: of	4-1¾"		" " connt rod 3¼ × 3¾ "			" " centres of —			

BOLIVIA RAILWAY (F. C. de B.)

In May 1906 a contract was signed between the Bolivian Government and the National City Bank and Speyer & Company (of New York) to provide for the construction and operation of railway lines in Bolivia. The two companies then promoted the organisation of a separate entity to be known as the Bolivia Railway Company and this was incorporated on 18 February 1907 and had assigned to it the above concession. Construction had started on a 128 mile (207km) line connecting Oruro and Viacha in 1906−this being opened to traffic during October 1908. Agreement was reached whereby the FCAB

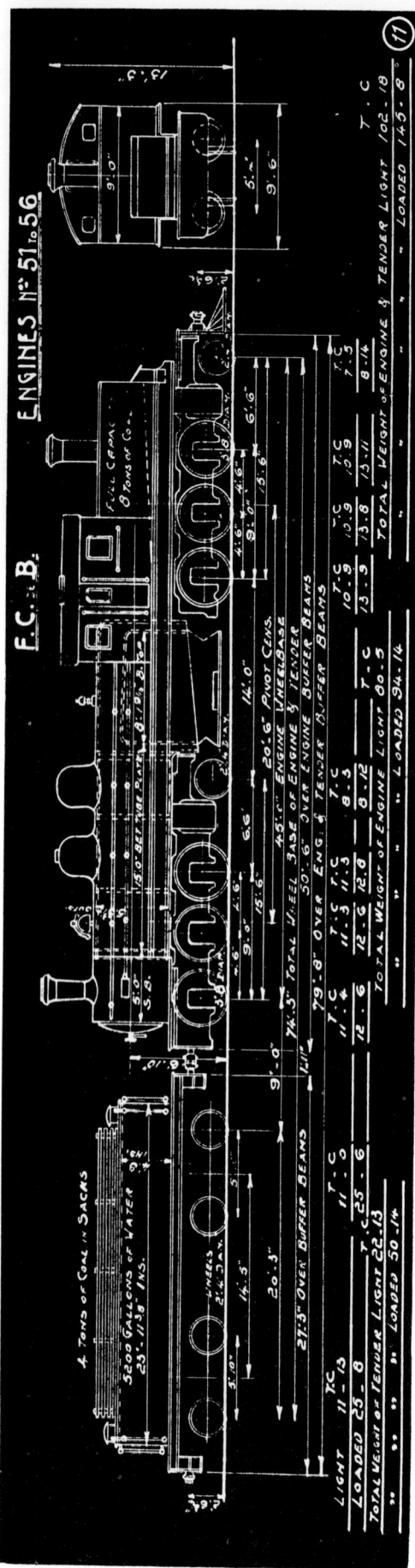

engineering contractor to the Bolivian Government but these lines could not be officially leased to the Bolivia Railway Co. until the Bolivian Government had satisfied the Bond-holders from whom the money to construct the lines had been borrowed. The Oruro–Viacha section was leased and operated by the FCAB from 1 January 1909.

The branch lines in Bolivia were also financed by the Bolivian Government and then leased to the Bolivia Railway Co. who sub-contracted with the FCAB for their operation. A 56 mile (90km) branch from Uyuni to Atocha was begun 1911, opened to traffic on 1 March 1913 and accepted under an FCAB operating lease from 1 July 1920. Work was inaugurated on the 108 mile (174km) branch from Rio Mulato to Potosi in 1909, being provisionally opened for services during 1912 and passing to the FCAB under an operating lease on 1 January 1916.

Work on the Viacha to La Paz section 20 miles (32km) to give access to "the highest capital in the world" was completed in 1917, under a concession granted in 1913. A further concession was then obtained and the line re-located, this work being completed in 1924.

Work started in 1909 on the 127 mile (205km) branch from Oruro to Cochabamba and this was completed by 26 July 1917. The FCAB operating lease was not activated until 1 January 1924.

Meyer type locomotives supplied to the Bolivia Railway
0–6–2+0–6–2 Beyer Peacock & Co. Ltd. Works No's 5617–22 of 1913 (ordered 1912). BP Order No. 0416. Type 2 simple locomotives with cylinders at the rear of each steam bogie and with separate tenders. FCAB Road No's 51–56, subsequently renumbered 451–6.

These six Meyer locomotives purchased by the F. C. de B. were most unusual to say the least, being of the 0–6–2+0–6–2 wheel arrangement and coming not from Kitson & Company but from the works of Beyer Peacock & Company in 1913. These were simple tank locomotives with the cylinders at the rear of each steam bogie and because of the high elevation of the Bolivian section, their cabs were fully enclosed, being arranged to run cab first. As in early Kitson Meyer locomotives two chimneys were provided, one in the normal position and a rear stack on the bunker to exhaust the rear steam bogie cylinders. No side tanks were provided but a bunker behind the cab held 8 tons of coal. A separate tender carrying 5,200 gallons of water and 4 tons of sacked coal rode on two 4-wheel bogies and was coupled to the smokebox end of the locomotive. This unusual arrangement required a crew of at least four, a driver, fireman and two coal carriers who were responsible for carrying the sacks of coal from the front end tender, along the footplate to the cab. How this dangerous and inconvenient arrangement came to be is not known, except that the specification called for the locomotives to be arranged to run cab first. After the unsuccessful Livesey Son & Henderson design for "Hercules", one might have expected the company to order rather more conventional motive power. Whether the Consulting Engineers or the FCAB themselves were responsible for the design is not known but one can imagine Kitson & Co. throwing up their hands in horror and refusing to build these monsters. We do not know if this was the reason why Beyer Peacock was given the order but at any event these were to prove really excellent locomotives.

From information provided by the Company it appears these

would lease the line from the Bolivia Railway Co., and from 1 January 1909 would take over its working. The FCAB constructed the railways in Bolivia in the capacity of a civil were built new to 1 metre gauge and in later years they were converted to oil burners. The Bolivian section of the F. C. Antofagasta (Chili) & Bolivia commenced from the Chilian border at Ollague, the main line proceeding to Uyuni, Oruro, Viacha and La Paz, 721 miles from Antofagasta. In later years these 6 Meyer locomotives were used between Oruro and Viacha and occasionally to La Paz on freights. They were also passed for service on the branch from Uyuni to Atocha and Tupiza, and from Rio Mulato to Potosi, also the Cochabamba branch from Oruro. The Rio Mulato—Potosi branch (170km) passed through rugged mountain country and included 1 in 30 grades uncompensated for 245ft radius curves. Along this branch was Condor, at 15,814ft the highest station in the world.

The Bolivian main line ran over undulating pampa country with easy grades and curves excepting for the last 9½ miles into La Paz where the line descended 919ft very rapidly by 1 in 33 grades and 15 degree curves.

In the period 1953—56 the Bolivian Railway (in effect the FCAB), had all six Meyer locomotives at work and they were still in service in 1959 when the Antofagasta (Chili) & Bolivia Railway Company ceased operations in Bolivia following Government nationalisation of the railways. By 1971 the Bolivian Railways were mainly diesel hauled and in 1976 these six 0−6−2+0−6−2 locomotives were dumped at the main Bolivian works at Uyuni. It is believed they were still there in 1989 along with some Garratt locomotives.

In 1923/4 the Beyer Peacock Meyer 0−6−2+0−6−2 were rated to haul:

Main Line:

Uyuni—Rio Mulato	670 tons
Rio Mulato—Uyuni	670 ,,
Rio Mulato—Oruro	792 ,,
Oruro—Rio Mulato	792 ,,

Oruro—Viacha:

Oruro and Chijmuni	900 ,,
Chijmuni and Oruro	1000 ,,
Chijmuni and Calamarca	800 ,,
Calamarca and Chijmuni	1000 ,,
Calamarca and Viacha	1000 ,,
Viacha and Calamarca	785 ,,

Viacha—La Paz:

Viacha and Kenko	790 ,,
Kenko and La Paz	800 ,,
La Paz and Kenko	220 ,,
Kenko and Viacha	1000 ,,

Ramal Tupiza:

Uyuni and Cerdas	700 ,,
Cerdas and Uyuni	750 ,,
Cerdas and Atocha	300 ,,
Atocha and Cerdas	230 ,,

Ramal Potosi:

Rio Mulato and Potosi	215 ,,
Potosi and Rio Mulato	215 ,,

Oruro—Cochabamba:

Oruro—Banderani	290 ,,
Banderani—Arque	290 ,,
Arque—Quillacollo	290 ,,
Quillacollo—Cochabamba	300 ,,
Cochamba—Buen Retiro	792 ,,
Buen Retiro—Tolapalca	215 ,,
Tolapalca—Oruro	792 ,,

Railway	FCAB
Gauge	1m
Wheels	0−6−2+0−6−2
Maker	Beyer Peacock
Works No.	5617−22
Year	1913
Cylinder Position	rear of each unit
Cylinders—inches	18 × 20
Boiler Pressure—lbs/sq in	160
Heating Surface:	
Firebox—sq ft	153
Tubes—sq ft	2120
Total—sq ft	2273
Superheater—sq ft	None
Total HS—sq ft	2273
Grate area—sq ft	40.8
Driving wheel—diameter	3ft 8in
Other wheels—diameter	2ft 4in
Rigid wheelbase	9ft 0in
Wheelbase—first group	15ft 6in
Wheelbase—second group	15ft 6in
Total wheelbase	45ft 0in
Water—galls	None
Coal—tons	8
Weight empty—tons/cwt	80 5
Weight in W.O.—tons/cwt	94 14
Weight Adhesive—tons/cwt	
Tractive effort—lbs	34,600
Height—overall	13ft 3in
Width — ,,	
Length— ,,	
Tender:	
Water—galls	5200
Coal—tons	4 in sacks
Wheel diameter	2ft 4in
Total length engine and tender	79ft 8in
Hauling capacity on straight level track @ 8−10mph—tons	
On 1 in 100 @ 8−10mph—tons	
On 1 in 75 @ 8−10mph—tons	
On 1 in 50 @ 8−10mph—tons	
On 1 in 25 @ 8−10mph—tons	

FCB No. 452 was built by Beyer Peacock & Co. Ltd as one of six Meyer type 0−6−2+0−6−2 supplied in 1913. When photographed in 1956 it had been converted to oil burning, the oil tank being located behind the cab. A separate water tank was provided at the smokebox end.
D. Ibbotson.

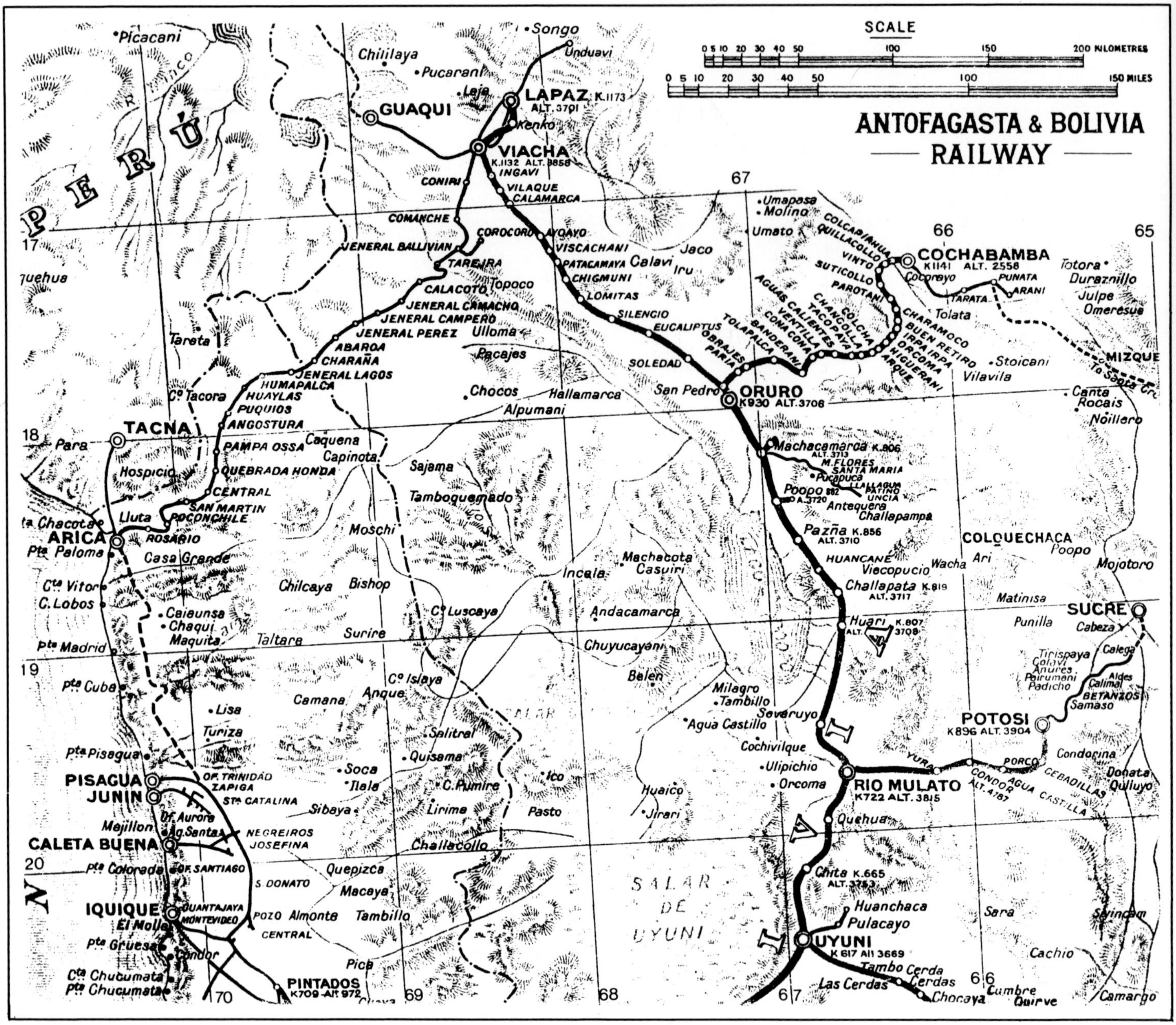

No. 455 was photographed in 1956 and still retained its rear stack. Detail differences include a central handwheel on the smokebox door in contrast to the photograph of No. 452 (on the previous page) which had its smokebox door secured by dogs. No. 452 also had a higher built-up oil bunker. No. 455 was minus its air compressor on the side shown whereas No. 452 had a compressor mounted on both sides of the smokebox.

D. Ibbotson.

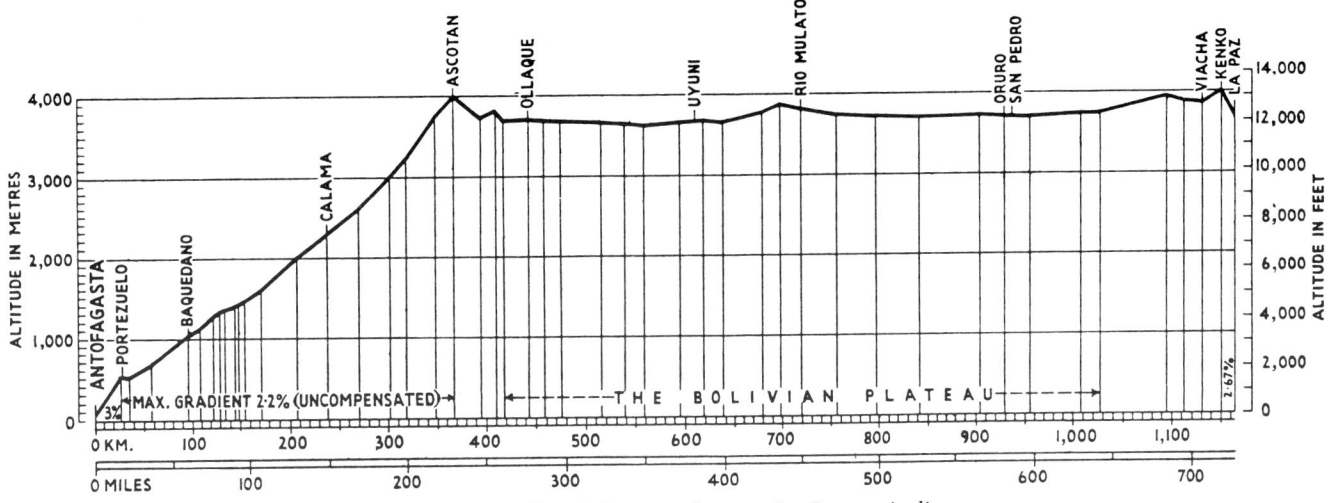

Gradient profile of the Antofagasta-La Paz main line

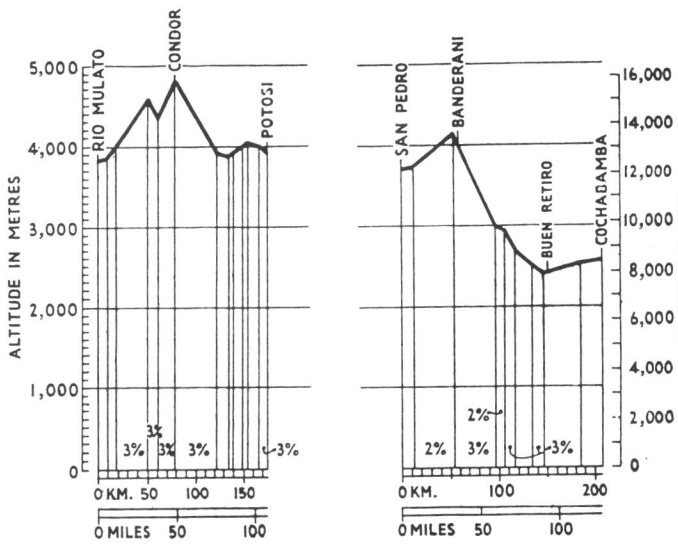

A Beyer Peacock 0−6−2+0−6−2 proceeds down a street in this 1956 view. Since these locomotives operated cab first, no smokebox headlamp was provided.　　　　　　　　　　　　　　　*D. Ibbotson.*

Potosi Branch　　　　　　　*Cochabamba Branch*

No. 452 at Oruro shed in 1956. Note the air compressors on each side of the smokebox.　　*D. Ibbotson.*

An unidentified Meyer 0−6−2+0−6−2 at work in Bolivia circa 1930. Note the high extension to the bunker, the twin air compressors and the re-railing jacks on the footplate. Note also the original long low front footplate. All six locomotives were later altered, their front footplates then assuming the appearance of No. 455. (see page 36).

D. Binns collection via M. Turner.

No 456 refills its auxilliary water tender in this 1956 photograph.

D. Ibbotson.

THE LEOPOLDINA RAILWAY

In the mid-1920s this 1 metre gauge Company was a British owned combination of 38 railways totalling some 1,854 route miles. Connecting Victoria with Rio de Janeiro, the Leopoldina served a rich sugar, rice and coffee district and ran through difficult country with heavy engineering works, 1 in 33 grades and 115ft radius curves.

Type 4 Livesey Son & Henderson Meyer 2−6+6−4T Kitson & Company, Works No. 4568/1908 for the Leopoldina Railway, Brazil.

D. Binns collection.

Kitson Meyer locomotives supplied to the Leopoldina Railway

The Company purchased one Livesey Son & Henderson type Meyer articulated for use on light track and steep gradients, but research has not shown where this was employed. Built to the specifications of the Consulting Engineers this was a type 4 simple locomotive with its cylinders at the outer ends of each bogie and having two steam bogies, the second supporting the back of the locomotive, the cab and tender front. The rear of

the tender was carried on an unpowered 4-wheel truck. Fittings included a steam brake operating on all coupled wheels together with fitting for the Eames vacuum brake on the train. An extended smokebox with a baffle plate was incorporated along with a wire mesh spark arrestor. 3,000 gallons of water was carried and 5 tons of briquettes. The maximum axle load was 9½ tons.

2−6+6−4T Kitson & Co. Works No. 4568 of 1908. Steam trial 10/11/1908. Type 4 simple locomotive with the cylinders at the outer ends of each bogie. Two powered trucks and one 2-wheel non-powered leading bogie and a 4-wheel non-power trailing truck. LR Road No. 139.

One of the difficulties for the researcher is the fact that two photographs exist of No. 139—one showing it with a rear chimney, the other without. This difference also occurs on pictures of the Gt. Southern of Spain 2−8+8−0T and one wonders why the rear chimney was touched out by the photographer.

The Leopoldina Railway purchased no further Kitson Meyer locomotives, but did try a Baldwin experimental 4+4 articulated in 1914, followed by Beyer Garratt's in 1929, 1937, 1940 and 1946.

Railway	Leopoldina
Gauge	1m
Wheels	2−6+6−4T
Maker	Kitson
Works No.	4568
Year	1908
Cylinder Position	outer ends
Cylinders—inches	14 × 18
Boiler Pressure—lbs/sq in	175
Heating Surface:	
Firebox—sq ft	107
Tubes—sq ft	1184
Total—sq ft	1291
Superheater—sq ft	None
Total HS—sq ft	1291
Grate area—sq ft	25
Driving wheel—diameter	2ft 10¾in
Other wheels—diameter	2ft 2in
Rigid wheelbase—each group	8ft 0in
Wheelbase—first group	13ft 7½in
Weelbase—second group	18ft 2in
Total wheelbase	42ft 2in
Water—galls	3000
Briquettes—tons	5
Weight in W.O.—tons/cwt	79 3
Weight Adhesive—tons/cwt	55 5
Tractive effort—lbs	26,643 (75%)
Height—overall	12ft 5¾in
Width — ˮ	8ft 0in
Length— ˮ	47ft 11in
Hauling capacity on straight level track @ 8−10mph	2981 tons
On 1 in 100 @ 8−10mph	770 tons
On 1 in 75 @ 8−10mph	605 tons
On 1 in 50 @ 8−10mph	414 tons
On 1 in 25 @ 8−10mph	189 tons

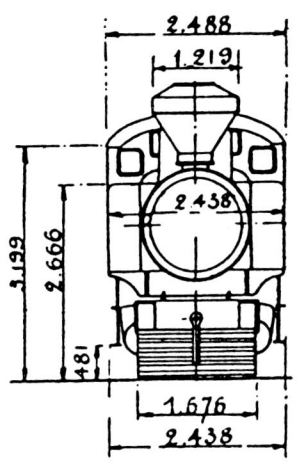

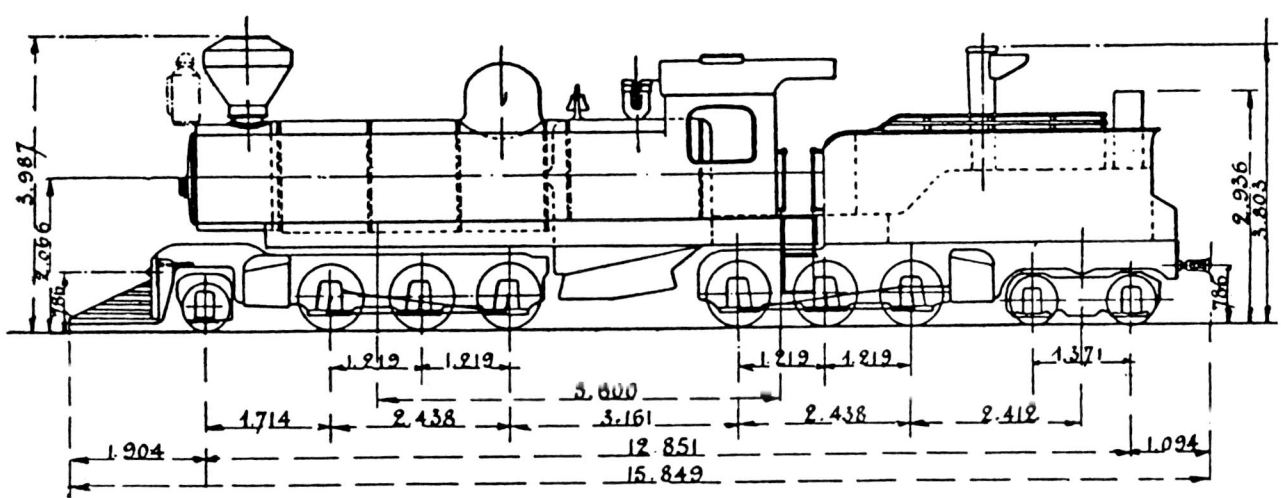

2−6+6−4T designed by Livesey Son & Henderson, Consulting Engineers to the Leopoldina Railway. This drawing is reproduced from "Articulated Locomotives" by Lionel Wiener (published by Constable & Co. 1930 and re-published 1970 by Kalmbach Publishing, USA). The drawing shows a diamond stack chimney indicating that this locomotive was a wood burner. A rear chimney is shown on the drawing but may have *been touched out on the photograph which is reproduced from a Kitson & Co. 1923 catalogue.*   *D. Binns collection.*

COLOMBIA:
GIRARDOT RAILWAY
CUNDINAMARCA RAILWAY

4

At 31 December 1925 the Ferrocarril de Girardot had 132 kilometres of main line track in operation and was Government owned and operated. The history of the F. C. de Girardot dates back to 21 June 1881 when Francisco J. Cisneros started construction of the first section, extending from the town of Girardot to Tocaima, a distance of 33 kilometres. Construction was supervised by a "junta directiva" created specially for this purpose. On 24 October 1884 Cisneros was authorised to extend the line from Tocaima to Juntas de Apulo, but this project was not completed due to the Colombian civil war of 1885. On 1 May 1886 the Colombian Government took over the railway and continued its construction to Juntas de Apulo, a distance of 39 kilometres. On 17 September 1887 the Government entered into a contract with Carlos Uribe to continue construction forward from Juntas de Apulo to the Sabana Railway. On 4 November following, this contract was transferred from Carlos Uribe to Francisco J. Cisneros. The Girardot Railway was managed by a syndicate until 23 September 1890 when a contract was legalised between the Colombian National Government and Don Luis Nieto, a representative of the River Plate Loan & Agency Company (of London). This arrangement was annulled in 1892 because the R.P.L.&A. Co. had not carried out its part of the contract. On 24 May 1892 the Government contracted with a John Pennington but declared this void in September 1894 because construction work had stopped for more than 6 months.

The Colombian Government again took over the Girardot Railway pending a new contract made on 26 December 1894 with Samuel B MacConnico under whose control an extension was begun from Anserma toward Anapoima.

On 25 March 1895 MacConnico transferred his rights to the Great Colombian Railway Co. who transferred their contract on 31 July 1897 to Jose Manuel Goenaga. On 3 January following, Goenaga transferred his contract to Juan B. Mainero y Truco.

After a chequered career things were finally about to move and on 19 June 1898 the Colombian National Railway Co. was incorporated in England to obtain a concession from the Government of Colombia for the exclusive right to build and operate a railway extending from Bogota to Girardot.

The Company took over the already completed line (29 kilometres) and works, and continued construction until 23 August 1909 when the Girardot Railway was opened over its whole length. The Company operated the line until 1922 but indebtedness to the Colombian Government increased each year until, in July 1922, an agreement was reached for the sale of the road to the Colombian Government—this agreement being approved by the stock holders of the Company in December 1922.

So it was that from 1898 to the end of 1922 the Girardot Railway (Girardot—Facatativá) was owned by the British Company "The Colombian National Railway Co." (418 Strand, London). In spite of the name this Company was independent from the Colombian Government. The C.N.R. Co. obviously had knowledge of activities in Chile judging from the fact that the engineer Alexander Gulliver came from the Transandine Railway to finish the road to Girardot (which he did in 1909 with the help of labour from the Government).

Although the 3ft 0in gauge (0.914 metre) Ferrocarril de Girardot opened on 23 August 1909, a correspondent in the "Engineer" for 11 March 1910 wrote: "I much question whether it can be regarded even now as fit to be open since it is far from finished and, at the best cannot be completed for several years to come."

The total length of main line was 132 kilometres from Girardot (1066ft above sea level) on the banks of the navigable Magdalena river—climbing to 8953ft near to the upper terminus at Facatativá where the Girardot Railway was to join up with the Sabana (Savannah) Railway to provide a continuous route from the coast, via the Magdalena river, to the Capital city of Bogota (the terminus of the Sabana Railway). Unfortunately little overall planning seems to have gone into the Colombian rail network, since the Girardot section was built to 3ft 0in gauge, and the Sabana Railway with which it connected, to 1 metre. This required a trans-shipment from one to the other—a task made even more difficult due to one train stopping more than 60ft behind the other and as a consequence every item had to be manhandled. Because of the time taken to tranship, 1 hour was lost out of a total journey of 12 hours.

The mountainous character of Colombia required a rise of some 5800ft in 33 miles with a large number of deep cuttings, one tunnel 510ft long located at km 115, and bridges over the numerous rivers. There were 53 bridges of which 6 were timber (with a total length of 151.8 metres), 24 masonry and timber (with a total length of 160.61 metres), and 23 of masonry and steel (total length 353.52 metres). The total length of bridgework was therefore 665.93 metres. Although not as mountainous as Bolivia, Chile or Peru, the climb on the Girardot Railway was mostly at 1in25, continuous for considerable distances, combined with almost continous curves of 260—300ft radius. Between km 70 and 125 from Girardot there was an average grade of 3.43% compensated, with a maximum grade for a distance of 40km of 4% compensated. The minimum curve at km 85 was 57 metres, this being uncompensated and there was almost no tangent between it and the next curve. The sharpest curves elsewhere had a radius of 70 metres but none were compensated. In the mid-1920s rails weighing 45lb and 60lb/yd were laid on ties (sleepers) spaced at 60cm, ballasted with rock. The roadbed was reported as being in excellent condition. Rather strangely for a railroad whose roadbed was in excellent shape, and whose locomotives were modern and up to date, the signalling system left much to be desired, the signalling equipment comprising coloured flags and lanterns.

STATIONS	Altitude metres	Water Station litres	Coal Stations tons
Girardot	326	24,000	40
La Virginia	336	–	–
Tocaima	390	10,000	10
Portillo	408	–	–
Juntas de Apulo	455	40,000	–
Anapoima	572	–	–
San Joaquin	656	15,000	–
La Mesa	963	–	–
Hospicio	1068	–	–
La Esperanza	1285	14,000	30½
Cachipay	1632	–	–
Anolaima	1972	12,000	10
Zipacon	2495	47,000	500
El Chuscal	2728	–	–
Facatativá	2614	–	5

Additional water stations were located at San Javier 34,000 litres, and at km 20 30,000 litres. Coal was also available at San Javier (20 tons capacity). Soft coal was used as fuel with an excellent grade of bituminous coal being found at Zipacon, km 120.

Both British and American engineering firms were responsible for the bridge construction (Matthew J. Shaw & Co., and H. J. Skelton & Co.—both of London, also the United States Steel Products Export Co., of New York). Rolling stock provided for the opening of the line was the work of American and Belgian manfacturers with 23 passenger and 3 luggage cars made by the American Car & Foundry Co. There were also 49 freight cars and platforms of which 10 were made in Belgium.

The locomotive stock obtained for the opening of the Girardot Railway was an odd assortment from British and American manufacturers and included products of the Rhode Island Locomotive Works, the Baldwin Locomotive Works and Alco, the latter furnishing three 2−8−0 tender engines, and two semi-articulated 0−6−6−0T Mallet types (Works No's 46167/8) these latter carrying road numbers 19 and 20 (later 9, 10). From England Hudswell Clarke & Co. (Leeds) supplied two locomotives and one came by way of Jones Burton & Co. (Liverpool). The principle power was supplied by Kitson & Co. in the shape of three Kitson Meyer 0−6+6−0T.

The Kitson Meyer and Mallet locomotives were used on the upper steeper sections between La Mesa 3153ft and Facatativá station, passing over the summit 9088ft. At the time of publication of my original book "Kitson Meyer Articulated Locomotives" (Wyvern Publications January 1985), there were many unanswered and seemingly unanswerable questions regarding the Girardot Kitson Meyer locomotives. However, subsequent and lengthy correspondence with Gustavo Arias-de-Grieff in Colombia, and in particular with P. K. Dewhurst (son of P. C. Dewhurst) in the USA, has cleared up many of the unanswered queries and I am deeply indebted to both these correspondents. What follows is a revised version of my original text incorporating all the additional information.

The first three Kitson Meyer 0−6+6−0T destined for the Girardot Railway emerged from the works of Kitson & Co. in 1909. These were followed by two more in 1912 and a further pair in 1914—all seven using saturated steam. Two additional 0−6+6−0T were delivered in 1914 and these and all subsequent locomotives were fitted with superheaters, Schmidt type

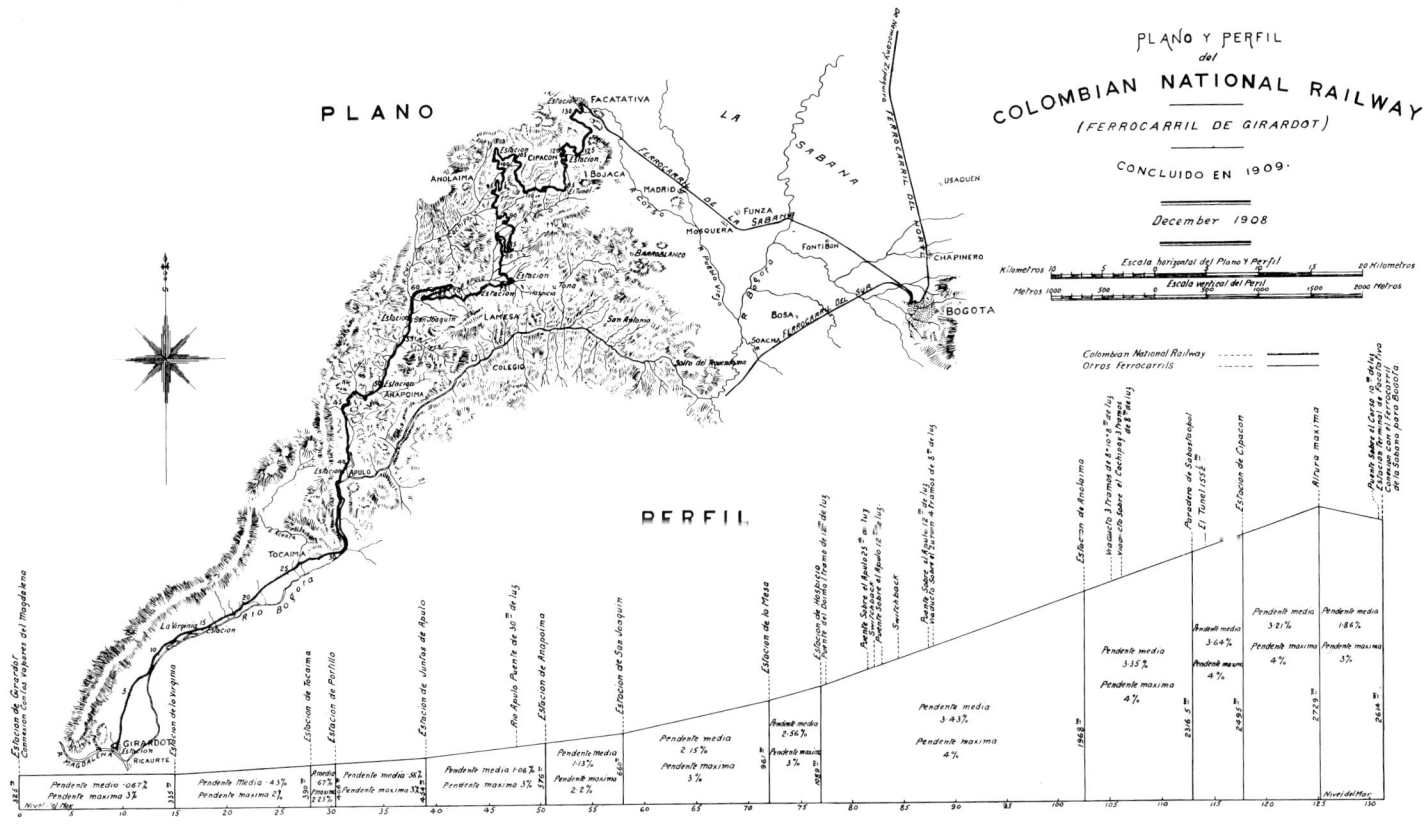

PLAN SHEWING CURVES AND GRADIENTS NEGOTIATED BY KITSON-MEYER ENGINE.

on the first two and Robinson type on all subsequent batches. A further three Kitson Meyer 0−6+6−0T were ordered in 1918 and shipped in the following year, and these were followed by three more in 1920 (two of which were exhibited at the Works of Kitson & Co. on the occasion of Lt. Col. Edwin Kitson Clark's paper to the Institute of Locomotive Engineers). The final four Kitson Meyer 0−6+6−0T emerged in 1920, making 19−all for use on the hill section between La Mesa and Facatativá on the F. C. de Girardot.

The subject of loco numbers was also one which had proved most vexing at the time the original research was carried out. Details of Road No's were extracted from Kitson & Co. steam trials book currently in the possession of the National Railway Museum at York. These numbers were inked in but the source of this information is not known. Because one of the re-numberings is shown as advised by letter, it was felt that these numbers must have come from official sources in Colombia. These were subsequently proved to be largely inaccurate and the true Road No's have been supplied by Gustavo Arias who obtained them from F.C.N. official roster sheets. The re-numbering from 1909-1929 numbers to those carried during the period 1930-1937 were confirmed by P. K. Dewhurst with new No. 1 being old No. 5−an Alco 2−8−0. New No. 2 was old No. 4−also an Alco 2−8−0. The 19 Kitson Meyer 0−6+6−0T were renumbered 3−21. The document appertaining to this renumbering is headed "Particulars of the Locomotives of the F. C. de Girardot y Tolima, Huila Caqueta" with the following note "As the Girardot Railway and the Tolima line are now **ONE** railway, the loco's have all been renumbered as belonging to one railway. I got out the new numbering with the idea of avoiding renumbering for at least ten years and this was approved by Dr. Escallon".

Facatativá station with 3ft 0in gauge Girardot Railway Type 1 Kitson Meyer 0−6+6−0T No. 21 (Kitson Works No. 5065/1914) at left. On the right, a locomotive belonging to the F.C. de la Sabana heads a train on this metre gauge railway. P. C. Dewhurst via P. K. Dewhurst.

Girardot Railway No. 18 (Kitson Works No. 5039/1914) was photographed at Girardot Works on 31 July 1924 in lined green livery but minus ownership markings. P. C. Dewhurst via P. K. Dewhurst.

F.C. Girardot Kitson Meyer 0–6+6–0T No. 12 was Kitson & Company, Works No. 4671/1909 and one of the first three supplied to Colombia. It was a Type 1 locomotive with cylinders at the rear of each steam bogie, saturated steam, and had a rear chimney. Note also the short front end frames (in front of the leading drivers).
via P. K. Dewhurst.

No. 15 was Kitson & Co. Works No. 4915/1912 and differed only in detail to No. 12 above. No. 15 had coal rails and a bell but no rear chimney (unless of course this was touched out on the official photograph). No. 12 (top) had a nice cast number plate whereas "No. 15" is painted on the tank side. Note the lifting holes at each end of No. 15's frame (missing on No. 12) and the lack of air reservoirs beneath the side tank behind the front unit cylinder.   via P. K. Dewhurst.

No. 29 was Kitson & Co. Works No. 5322/1920 and was provided with a Robinson superheater when new. In appearance it was virtually the same as No. 15 pictured above.
via P. K. Dewhurst.

KITSON MEYER TYPE 0−6+6−0T DELIVERED TO THE F.C. de GIRARDOT

Note: All numbers shown are F.C. Girardot numbers up to 1931 after which they are for F.C. Girardot−Tolima−Huila, until the formation of the Centrales Division of the National Railways in 1953.

Kitson No.	Date Built	Steam Trials	No. as Built	1909 to 1929	1930 to 1931	1931 to 1937	1938 to 1945	1946 to 1953	1953 on	Notes	
4671	1909	14/9/09	12 ★	12	3	3	9	–	–	A	Saturated steam when built
4672	1909	9/9/09	13 ★	13	4	4	10	16	141	B	Saturated steam when built
4673	1909	9/9/09	14 ★	14	5	5	11	–	–	A	Saturated steam when built

★ Altered by letter 6/1911 to 9, 10, 11. This is unlikely and P. K. Dewhurst is sure these numbers were never carried. No's 9 and 10 were already occupied by two Alco Mallet 0−6+6−0Ts (built as No's 20 and 21) and No. 11 was an 0−4−0ST when the first KM's arrived. Also the official photo of No. 12 taken in finished state shows a cast number plate.

Note: A. One out of either 4671 or 4673 went to F.C. Cundinamarca after 1946 to become their No. 14, the other seems to have disappeared.

Note: B. Figures in Chart #461, FCN 1957, with C/N.
    Figures as retired in Chart February 1954, with No. 16.
    No Construction No.

Kitson No.	Date Built	Steam Trials	No. as Built	1909 to 1929	1930 to 1931	1931 to 1937	1938 to 1945	1946 to 1953	1953 on	Notes	
4915	1912	21/10/12	15	15	6	6	–	–	–	C	Saturated steam when built
4916	1912	25/10/12	16	16	7	7	21	21	146	D △	Saturated steam when built

Note: C. Not in 1938 listing−According to P.C. Dewhurst's notes made in the 1930s this loco was sold to or taken by the Army Engineers about 1934 for construction use on the Cundinamarca below Facatativá, it was not returned to the Girardot.

Note: D. Figures as retired in Chart #318, FCN February 1954, without C/N, built 1914.

Note: △ Later fitted with 180lb (type D boilers) but front truck not rotated. Wheelbase 318in (per 1946 roster).

Kitson No.	Date Built	Steam Trials	No. as Built	1909 to 1929	1930 to 1931	1931 to 1937	1938 to 1945	1946 to 1953	1953 on	Notes	
5039	1914	13/6/14	18	18	8	8	22	22	147	E △	Saturated steam when built
5040	1914	20/6/14	19	19	9	9	23	23	148	F △	Saturated steam when built

Note: E. Figures as retired in Chart #318, FCN February 1954, without C/N.

Note: F. Figures as retired in Chart #318, FCN February 1954, without C/N.

Note: △ Later fitted with 180lb (type D boilers) but front truck not rotated. Wheelbase 318in (per 1946 roster).

Kitson No.	Date Built	Steam Trials	No. as Built	1909 to 1929	1930 to 1931	1931 to 1937	1938 to 1945	1946 to 1953	1953 on	Notes	
5064	1914	20/8/14	20	20	10	10	12	–	–		Schmidt sup'htr when built
5065	1914	21/8/14	21	21	11	11	24	24	149	G △	Schmidt sup'htr when built

Note: G. Figures as retired in Chart February 1954, without C/N.
    Figures in Chart #461, FCN 1957, with C/N

Note: △ Later fitted with new trailing bogie at front end to become Pseudo type 3.

Kitson No.	Date Built	Steam Trials	No. as Built	1909 to 1929	1930 to 1931	1931 to 1937	1938 to 1945	1946 to 1953	1953 on	Notes	
5176	1919	27/6/19	23	23	12	12	13	–	–		Robinson sup'htr when built
5177	1919	7/7/19	24	24	13	13	14	–	–		Robinson sup'htr when built
5178	1919	18/7/19	25	25	14	14	25	25	150	H △	Robinson sup'htr when built

Note: H. Figures as retired in Chart #318, FCN February 1954, without C/N.

Note: △ Later fitted with new trailing bogie at front end to become Pseudo type 3.

Kitson No.	Date Built	Steam Trials	No. as Built	1909 to 1929	1930 to 1931	1931 to 1937	1938 to 1945	1946 to 1953	1953 on	Notes	
5274	1920	6/9/20	26	26	15	15	15	20	145	I	Robinson sup'htr when built
5275	1920	14/9/20	27	27	16	16	–	–	–	J	Robinson sup'htr when built
5276	1920	18/9/20	28	28	17	17	16	17	142	K	Robinson sup'htr when built

Note: I. Figures as retired in Chart #318, FCN February 1954, without C/N.

Note: J. Not in 1938 Listing. According to P.C.D. notes made in 1930s, 5275 was sold to or taken by the Army Engineers about 1934 for construction use on the Cundinamarca below Facatativá, it did not come back to the Girardot.

Note: K. Figures as retired in Chart #318, FCN February 1954, without C/N.
    Shown as built 1914.

Kitson No.	Date Built	Steam Trials	No. as Built	1909 to 1929	1930 to 1931	1931 to 1937	1938 to 1945	1946 to 1953	1953 on	Notes	
5322	1920	27/ 9/20	29	29	18	18-19	20	–	–	L	Robinson sup'htr when built
5323	1920	6/10/20	30	30	19	19-18	17	–	–	L	Robinson sup'htr when built
5324	1920	11/10/20	31	31	20	20	18	18	143	M	Robinson sup'htr when built
5325	1920	21/10/20	32	32	21	21	19	19	144	N	Robinson sup'htr when built

Note: L. According to P.C.D's notes, in the mid 1930s 5322 running as No. 18 was loaned to the F.C. Cundinamarca and later came back as No. 19.
    5323 formerly running as No. 19 also loaned and came back as 18.

Note: M. Figures as retired in Chart #318, FCN February 1954, without C/N.

Note: N. Figures as retired in Chart, February 1954, without C/N.
    Figures with C/N in Chart #461, FCN 1957.
    Photo shows legend 'Centro de Ingenieros Militares Francisco J. Caldas'−used for training purposes by military engineering section of army.

All the above were simple locomotives built new with cylinders at the rear of each steam bogie and at this stage were very much the standard Kitson & Co. product. By 1930 some of the earlier saturated steam 0−6+6−0Ts had received superheated boilers and all continued to give good service under extremely difficult operating conditions.

At the time of publication of my book it had not been possible to obtain photographs of all the different batches of Girardot 0−6+6−0Ts and it was thought that a change over had been effected in the position of the cylinders with the 1912 delivery, the cylinders being relocated from the rear of each bogie to the outer end of each bogie. But if this was the case how could a photograph of a 1920 locomotive (No. 29)−which appeared in Kitson & Co's 1923 catalog with its cylinders at the rear of each steam bogie−be explained? First thoughts were that Kitson & Co. had used a suitably renumbered photograph of an earlier locomotive to illustrate No. 29 and for some time this was considered a likely explanation. At the same time doubt existed with regard to No. 25, a photograph showing this locomotive carrying 1920 Works plates and with its cylinders at the outer ends of each steam bogie. From the list it will be seen that No. 25 was completed in 1919 (Works No. 5178) and there remains the unanswered query as to why this locomotive caried a 1920 Works plate. Possibly the Road number was crossed with another locomotive in the shops at the same time, and No. 25 in the photograph was actually one of the 1920 units. Or, perhaps the wrong Works plate was attached following a repair. The main problem was that No. 25, which had been built before No. 29, was to the more modern configuration with its cylinders at the outer ends of each unit. Having supplied this type of locomotive it seems unlikely that later deliveries would revert to the original configuration.

This question of cylinder positioning was put to both correspondents, P. K. Dewhurst replying "I recall never ever having seen a Girardot 0−6+6−0T with the cylinders on the leading truck towards the front . . . they were all with cylinders at the rear of each bogie. I am quite certain of this and I must have seen them all at some time or other. Furthermore if any had been converted prior to April 1928 when we left Colombia

P.C.D. would certainly have photographed them." P. K. Dewhurst went on to say "By the way, my memory tells me that there is a letter somewhere showing that P.C.D. authorised the turning round of the front bogie on two 0−6+6−0Ts towards the end of 1928 but I am certain that the work was not done before April 1929. I'm sure it was done to reduce flange wear, I will try to locate the evidence."

Further correspondence revealed a letter from Seth R. Phelps to P. C. Dewhurst dated 1 December 1928 indicating that Phelps had by then ordered two sets of trailing bogies to fit up a pair of small Kitson Meyer's with the leading bogie the other way round. Phelps was Works Manager at Girardot and came under P. C. Dewhurst, so evidently P.C.D. had done the design work in Bogota and asked Phelps to order them through Girardot. (A letter dated 5 June 1934 from P.C.D.−then in Uruguay,−to Lt. Col. Edwin Kitson Clarke confirms that P.C.D. did the design work in Bogota in 1928/9). In fact blue print arrangement drawings show that P. C. Dewhurst was working on the rotation of the leading bogie on small Kitson Meyer's as far back as 1927. The drawings also give some idea on to why it was easier to use a rear bogie at the front rather than rotate the existing front bogie. Evidently it was due to the run of both live steam and exhaust pipes with their expansion arrangements.

The two new sets of trailing bogies were used on original No. 25 (Works No. 5178) and original No. 21 (Works No. 5065). All 19 Girardot 0−6+6−0Ts were delivered new with their cylinders at the rear of each unit and only the above two have been confirmed as having their front bogies reversed by using new trailing bogies (see 1938 Girardot roster), at the same time being fitted with the new larger type D boilers.

This is confirmed by a letter from P. C. Dewhurst to Lt. Col. Edwin Kitson Clarke dated 5 June 1934 from which the following is extracted: "Am very glad to hear Airedale is carrying out work for Colombia again . . . I see you are making a lot of parts etc. for rebuilt Kitson Meyer's; I wonder if this means the rebuilding of the 0−6+6−0T with the front bogie reversed as I had re-designed them in 1928/9 etc . . ."

Although the small Kitson Meyer's did an excellent job in Colombia, due to axle load limits the boilers were really too small for the cylinders (rather like the Jamaican KMs) and consequently the locomotives were heavy on boiler repairs. Then when P. C. Dewhurst got some spare time from the prime job of setting up the standard Colombian new locomotive designs, he turned to trying to reduce maintenance costs on the 19 Kitson Meyer 0−6+6−0Ts which he (correctly) felt had many more years of life in them. There were four types of boilers used with the 19 0−6+6−0T. These are detailed below:

dia. is a very good engine, shows about 5% saving in coal and pressure between 175 and 180 lbs on stiffest grades, best performance 145 tons between La Mesa and Facatativá at good speed.

No. 8 with blast pipe 4in below centre line of boiler and 4¼in dia. and a D & M Exhaust Injector is a marvel, back pressure in cyls. much less, saving in fuel between 10% and 15%, has taken a freight train of 150 tons at speed, and a passenger train

BOILERS FOR SMALL KITSON MEYER LOCOMOTIVES FOR F.C. de GIRARDOT

Boiler type	A.	B.	C.	D.	E.
Barrel O.D. (front end)	4ft 5⅞in	4ft 5⅞in	4ft 5⅞in	4ft 10¼in	4ft 5⅞in
Length (between tube plates)	11ft 0½in	11ft 0½in	11ft 0½in	11ft 8⅞in	11ft 0½in
Firebox length (outside)	7ft 1in	7ft 6in	7ft 6in	7ft 6in	7ft 6in
Firebox width (outside)	4ft 5in	4ft 5in	4ft 5in	4ft 5in	4ft 5in
Tubes O.D.	1¾in	1¾in	1¾in	1¾in	1¾in
Number	206	206	109	110	109
Flues O.D.	−	−	5¼in	5⅜in	5⅜in
Number	−	−	14	18	14
Heating surface					
Tubes sq ft	1042	1042	551.5	591.5	551.5
Flues sq ft	−	−	212.5	297.5	217.5
Firebox sq in	102	106.5	106.5	113	106.5
Total Evaporative	1144	1148.5	870.5	1002	875.5
Superheater	−	−	171	236	160.5
Total sq ft	1144	1148.5	1041.5	1238	1036
Grate area (sq ft)	25.5	27	27	27	27
Boiler pressure (lbs/sq in)	160	160	160	180	160

In 1923 the following boilers were fitted:

Type A Works No's 4671-3 Non Sup.

Type B Works No's 4915-6 Non Sup.
    Works No's 5039-40 Non Sup.

Type C Works No's 5064-5 Superheated
    Works No's 5176-8 Superheated
    Works No's 5274-6 Superheated
    Works No's 5322-5 Superheated

Type D was a later development used on the two locomotives which had their front trucks rotated. Also used on three others which did **not** have their front trucks rotated.

Type E was used on the pair of Cundinamarca Rly small 2−6+6−2Ts of 1928, Works No's 5416/7.

The following details are extracted from correspondence from Newbold to PCD (engines are referred to by the then new numbers−period 1930-7):

No's 4-6 and 9 had non superheated boilers.

No's 7-8 and 9 are rebuilt with big boilers.

No. 8 has exhaust injector and 4¼in dia. blast pipe 4in below centre line of boiler.

No. 7 has two ordinary injectors and 4¼in dia. blast pipe 4in below centre line of boiler.

No. 9 has two ordinary injectors and blast pipe as sent by Kitsons 4in dia.

No's 4 & 6 were to be rebuilt. I started No. 4 but was stopped.

No's 3-5 & 17 have the smallest boilers.

A letter from Newbold to P.C.D. in Uruguay dated 19 April 1932 gives further insight into the three locomotives then fitted with P.C.D's "big boilers" but retaining their original trucks:

No. 7 with blast pipe 4in below centre line of boiler and 4¼in

well filled of eight new stock coaches, she is a beauty.

No. 9. Medina apparently got jealous of the success of No. 9 and started telling tales up in Bogotá for I get instructions that I was not to introduce any improvements but to erect No. 9 as Kitsons had sent it, so I could not lower the Blast pipe or alter the diameter, the result is she loses pressure and has stalled two or three times, and eats coal like the devil."

Newbold was a relatively junior man when P.C.D. was in Colombia. He was sent down to assist Phelps some time after P.C.D. left. Phelps became Loco. Superintendent of the Tolima division before 1932.

Of particular interest is the larger version of the boiler for the small Kitson Meyers made in conjunction with Kitson drawing 1/27 dated 3 January 1927 which is listed in the table as type D. Clearly, this boiler which was about 8½in longer than the standard (as well as being larger in diameter) was intended to go with the "rotated bogie type." In Phelps's letter dated 1 December 1928 he notes "start on No. 16 with the new boiler . . . then proceed with No. 18, change that boiler, then No. 19 for 'cambios' (change) etc. . . ." This refers to Kitson Works No's 4916, 5039 and 5040 which received the new enlarged boilers but retained their original trucks. A re-examination of the boiler drawings showed that the 180 lb/sq in boiler had their "centre-line above rail" raised by 3in from 6ft 1in to 6ft 4in. This is conjunction with the 4⅜in increase in the diameter caused the top of the boiler to be just about level with the top of the side tanks instead of about 5½in below as was standard. Thus we have a visual means of identifying the large boiler. Another is to see if the locomotive had P. C. Dewhurst's standard top-feed arrangement between the chimney and dome. The new boilers also had P.C.D's standard 5⅜in o.d. flues. The domes on the new boilers were the same size as on the small boilers but would appear larger because all of it projected above the side tanks.

Another point clarified by P. K. Dewhurst was P.C.D's reference to the first three Kitson Meyer 0−6+6−0T as having short leading bogies. This was not a shorter wheelbase bogie; the reference meaning short front bogie frames, i.e. short frames in front of the first axle−see photograph of No. 12.

Total wheelbase over outer traction wheels.

Type W relates to Works No's 4671 to 4673.
   Dimensions 26ft 1in (313in).

Type X relates to all other 0−6+6−0 KMs when built.
   Dimensions 26ft 6in (318in).

Type Y relates to the locos which had the leading bogie rotated 180 degrees.   Dimensions 27ft 3in (327in).

Type Z relates to the two 2−6+6−2 KMs for Cundinamarca, Works No's 5416 and 5417.
   Dimensions 28ft 4in (340in).

All bogie units had the minimum 6ft 2½in (74.5in) rigid wheelbase.

Girardot Railway No. 21 (Kitson & Co. Works No. 4916/1912) at Girardot station circa 1960 with its steam dome uncovered.
G. Diaz via G. Arias.

Girardot Railway No. 30 (Kitson & Co. Works No. 5323/1920) was photographed at Girardot on 25 March 1923.
P. C. Dewhurst via P. K. Dewhurst.

F.C. de GIRARDOT
0−6+6−0T   1938 ROSTER−OFFICIAL

Kitson	Built as Type	Cyls.	Cyls.	D.W.	Rigid W.B.	Total W.B.	BP	Boiler dia.	Firebox length	width	Tubes	H.S.	Sup. H.S.	Grate Area	Water	Coal lbs	Total wt. lbs	T.E.
4671	1	4	14×18	2'10¾"	6'2½"	26'1"	160	4'4"	7'1"	4'5"	109×1¾"×11'2", 14×5¼"	874½	171	25½	1720	6614	141680	27600
4672	1	4	"	"	"	"	"	"	"	"	"	"	"	"	"	"	"	"
4673	1	4	"	"	"	"	"	"	"	"	"	"	"	"	"	"	"	"
4915	Disposed of before 1938																	
4916	1	4	"	"	"	27'3"★	180	4'8"	7'6"	"	110×1¾×12'1", 18×5¼"	874½	171	27	"	"	145264	31066
5039	1	4	"	"	"	"★	"	"	"	"	"	"	"	"	"	"	"	"
5040	1	4	"	"	"	"★	"	"	"	"	"	"	"	"	"	"	"	"
5064	1	4	"	"	"	26'1"	160	4'4"	7'1"	"	109×1¾"×11'2", 14×5¼"	"	"	25½	"	"	141680	27600
5065	1	4	"	"	"	27'3"★	180	4'8"	7'6"	"	110×1¾×12'1", 18×5¼"	"	"	27	"	"	145264	31066
5176	1	4	"	"	"	26'1"	160	4'4"	7'1"	"	109×1¾"×11'2", 14×5¼"	"	"	25½	"	"	141680	27600
5177	1	4	"	"	"	"	"	"	"	"	"	"	"	"	"	"	"	"
5178	1	4	"	"	"	27'3"★	180	4'8"	7'6"	"	110×1¾×12'1", 18×5¼"	"	"	27	"	"	145264	31066
5274	1	4	"	"	"	26'1"	160	4'4"	7'1"	"	109×1¾"×11'2", 14×5¼"	"	"	25½	"	"	141680	27600
5275	Disposed of before 1938																	
5276	1	4	"	"	"	26'1"	160	4'4"	7'1"	"	109×1¾"×11'2", 14×5¼"	874½	171	25½	"	"	141680	27600
5322	1	4	"	"	"	"	"	"	"	"	"	"	"	"	"	"	"	"
5323	1	4	"	"	"	"	"	"	"	"	"	"	"	"	"	"	"	"
5324	1	4	"	"	"	"	"	"	"	"	"	"	"	"	"	"	"	"
5325	1	4	"	"	"	"	"	"	"	"	"	"	"	"	"	"	"	"

★By 1938 two locomotives (Works No's 5065 and 5178) had been altered to type three configuration, with type D boilers. Three others (Works No's 4916, 5039 and 5040) had received type D boilers but did not have their front trucks rotated− remaining type 1.

The Girardot Railway of the period is recalled by Peter K. Dewhurst in a letter to me dated 18 April 1986: "I can recall the scene quite easily. The up train from Girardot backing into La Esperanza station (1285ft above sea level, between Hospicio and Anolaima), engine at rear with the bunker towards the last coach. It would stop at the station and wait (about ½ hour) until the down train from Facatativá arrived in, also being pushed from the rear by the loco but usually with the engine the other way round, i.e. with the chimney nearest the last coach. In the meantime women would descend on the train from all directions with baskets of cooked chicken, vegetables, rice, fruits of all kinds etc., all balanced on their heads. These would be offered to the passengers through the windows. I would mention that the Girardot line had one double "switch-back". La Esperanza station with its railway hotel was situated in the middle of the section between the two switches. For this

One of a pair of Alco-built 0−6+6−0T Mallet tanks (Works No's 46167/8) supplied as a comparison to the Kitson Meyer type on the F.C. de Girardot. These locomotives had 13in × 20/20½in × 20 cylinders, 3ft 2in driving wheels, 200lb BP and 23400lbs maximum tractive effort. They were not flexible enough for service on the curving Girardot line and no further Mallet type locomotives were ordered.   *D. Binns collection.*

reason the trains in either direction always came into Esperanza station backwards with the Kitson Meyer's pushing from the rear.

Shortly after the arrival of the down train, the up train would go on up to Bogota and very shortly after that, the down train would proceed to Girardot:

. . . I can recall the scene as though it were yesterday. After all, the arrival of the train was La Esperanza's big moment of the day, passengers, mail, some baggage being loaded and unloaded, people greeting each other and exchanging news . . . The engines were painted green with F. C. de G. on the side tanks. They were beautifully kept, all polished and oiled with the domes (particularly) gleaming in the sun with the copper cap to the chimney shining.

The timetable reproduced below is dated 28 February 1923 at a time when only the small Kitson Meyer locomotives were in service. It is interesting to note that the total time taken to come up from Girardot was 7hrs 10mins, and allowing for the 96mins. in stations, leaves a net time in motion of 5hrs 34mins for the 132km. This works out at an average speed (start to stop between stations) of 23.7km (about 15mph) for the up trip. The timing allowed 3min per km (12.5mph) on the tough section and 2min per km (about 19mph) as far as San Joaquin. The speed on the down trip was scheduled exactly the same as the up trip. Unfortunately we have not been able to locate a schedule for the times when the 2−6+6−2T were in use.

FERROCARRIL DE GIRARDOT
ITINERARIOS
TRENES DE PASAJEROS
Subida

ESTACIONES	Llegada H.M.	Demora M.	Salida H.M.	Kilometros de Estacion a Estacion	Minutos por Kilometro
Girardot	−	−	7.20	−	−
La Virginia	7.50	3	7.53	15	2
Tocaima	8.19	8	8.27	13	2
Portillo	8.33	3	8.36	3	2
Juntas de Apulo	8.52	4	8.56	8	2
Anapoima	9.20	3	9.23	12	2
San Joaquín	9.37	8	9.45	7	2
La Mesa	10.20	4	10.24	14	2½
Hospicio	10.36	3	10.39	5	2½
La Esperanza	10.57	30	11.27	6	3
La Capilla	11.48	3	11.51	7	3
Cachipay	12.01	5	12.06	3	3½
Anolaima	12.41	10	12.51	10	3½
Zipacón	1.36	12	1.48	15	3
Facativá	2.30	−	−	14	3

Bajada

ESTACIONES	Llegada H.M.	Demora M.	Salida H.M.	Kilometros de Estacion a Estacion	Minutos por Kilometro
Facativá	−	−	8.50	−	−
Zipacón	9.22	10	9.32	14	3
Anolaima	10.17	5	10.22	15	3
Cachipay	10.52	3	10.55	10	3
La Capilla	11.04	2	11.06	3	3
La Esperanza	11.27	4	11.31	7	3
Hospicio	11.49	3	11.52	6	3
La Mesa	12.07	30	12.37	5	3
San Joaquín	1.12	8	1.20	14	2½
Anapoima	1.34	3	1.37	7	2
Juntas de Apulo	2.01	4	2.05	12	2
Portillo	2.21	2	2.23	8	2
Tocaima	2.29	8	2.37	3	2
La Virginia	3.03	2	3.05	13	2
Girardot	3.35	−	−	15	2

Este itinerario regira desde el 5 de marzo próximo.
Bogotá, febrero 28 de 1923.
El Secretario,
Manuel Abello H.

No. 30 was photographed on 20 September 1923 following an accident at km 108.   *P. C. Dewhurst via P. K. Dewhurst.*

Actual withdrawal dates are unkown for the 0−6+6−0T. (The first Diesels for Antioquia and Pacifico arrived in 1953; then in 1958 arrived another group, for Centrales and Pacifico, and this was the beginning of the end for steam). As it happened when the Diesels arrived, the steam engines were retired from service, but the official documents "decommissioning" them and authorising their disposal were not "legalised" at the same time. Gustavo Arias has in his possession a letter dated 1 March 1977, in which the Accounting & Statistics Directors requests that some locomotives that are out of service, be discharged from the assets of the Centrales Divisions. Among the locomotives noted were No's 141, 143, 145, 150−all Kitson Meyer 0−6+6−0T, plus No. 155 (old Cundinamarca Railway 2−6+6−2T No. 14. These had all been retired by 1964 when Gustavo Arias was Chief Mechanical Officer.

Two Girardot Railway 0−6+6−0T were taken over or sold to the Army Engineers about 1934 for use in construction of the Cundinamarca Railway below Facativá. Neither of these locomotives (Works No's 4915/1912 and 5275/1920) were returned to the Girardot Railway but it is not known if they went into CR stock, or remained with the army.

It is a pity that no Colombian Kitson Meyer was preserved.

This however was in a way a tribute to British quality and workmanship; their scrap was in a higher demand than that coming from other countries locomotives.

Ex-No. 25, Centrales No. 150 at Flandes circa 1960 (Kitson & Co. Works No. 5178/1919). This locomotive was a Dewhurst pseudo Type 3 with cylinders at the outer ends of each unit, and a larger boiler.
G. Diaz.

DIMENSIONS FOR THE 19 SMALL 0−6+6−0Ts AS BUILT

Railway	Colombian National	Colombian National	Colombian National	Colombian National	Colombian National
Gauge	3ft. 0in.	3ft. 0in.	3ft. 0in.	3ft. 0in.	3ft. 0in.
Wheels	0−6+6−0T	0−6+6−0T	0−6+6−0T	0−6+6−0T	0−6+6−0T
Maker	Kitson	Kitson	Kitson	Kitson	Kitson
Works No.	4671/2/3	4915/6	5039/40	5064/5	5176/7/8
Year	1909	1912	1914	1914	1918
Cylinder position	rear				
Cylinders – inches	14 × 18	14 × 18	14 × 18	14 × 18	14 × 18
Boiler Pressure – lbs./sq.in.	160	160	160	160	160
Heating Surface					
Firebox – sq.ft.	102	106.5	106.5	106.5	106.5
Tubes – sq.ft.	1042	1042	1042	764	764
Total – sq.ft.	1144	1148.5	1148.5	870.5	870.5
Superheater – sq.ft.	None	None	None	171	171
Total HS – sq.ft.	1144	1148.5	1148.5	1041.5	1041.5
Grate area – sq.ft.	25.5	27	27	27	27
Driving wheel – diameter	2ft. 10¾in.	2ft. 10¾in.	2ft. 10¾in.	2ft. 10¾in.	2ft. 10¾in.
Other wheels	None	None	None	None	None
Rigid wheelbase – first group	6ft. 2½in.	6ft. 2½in.	6ft. 2½in.	6ft. 2½in.	6ft. 2½in.
Rigid wheelbase – second group	6ft. 2½in.	6ft. 2½in.	6ft. 2½in.	6ft. 2½in.	6ft. 2½in.
Total wheelbase	26ft. 1in.	26ft. 6in.	26ft. 6in.	26ft. 6in.	26ft. 6in.
Water – galls.	1720		1839	1839	1839
Coal – tons	2½	2½	2½	2½	2½
Weight empty – tons cwt.	47 2				
Weight in working order – tons cwt.	60 19	63 5	63 5	64 4	64 4
Weight adhesive – tons cwt.	60 19	63 5	63 5	64 4	64 4
Tractive effort – lbs.	23,848 (75%)	23,848 (75%)	23,848 (75%)	23,848 (75%)	23,848 (75%)
Overall height	12ft. 1⅝in.				
„ width	8ft. 4in.				
„ length	37ft. 11in.				
Hauling capacity on straight track @ 8−10mph – tons	2710				
On 1 in 100 @ 8−10mph	708				
On 1 in 75 @ 8−10mph	559				
On 1 in 50 @ 8−10mph	386				
On 1 in 25 @ 8−10mph	182				

Railway	Colombian National	Colombian National
Gauge	3 ft. 0 in.	3 ft. 0 in.
Wheels	0-6+6-0T	0-6+6-0T
Maker	Kitson	Kitson
Works No.	5274/5/6	5322-5
Year	1920	1920
Cylinder position		
Cylinders – inches	14 × 18	14 × 18
Boiler Pressure – lbs./sq. in.	160	160
Heating Surface		
Firebox – sq. ft.	106.5	106.5
Tubes – sq. ft.	764	764
Total – sq. ft.	870.5	870.5
Superheater – sq. ft.	171	171
Total HS – sq. ft.	1041.5	1041.5
Grate area – sq. ft.	27	27
Driving wheel – diameter	2 ft. 10¾ in.	2 ft. 10¾ in.
Other wheels	None	None
Rigid wheelbase – first group	6 ft. 2½ in.	6 ft. 2½ in.
Rigid wheelbase – second group	6 ft. 2½ in.	6 ft. 2½ in.
Total wheelbase	26 ft. 6 in.	26 ft. 6 in.
Water – galls.	1839	1839
Coal – tons	2½	2½
Weight empty – tons cwt.		50 4
Weight in working order – tons cwt	64 4	64 4
Weight adhesive – tons cwt.	64 4	64 4
Tractive effort – lbs.	23,848 (75 %)	23,848 (75 %) 27,000 (85 %)
Overall – height		
,, – width		
,, – length		
Hauling capacity on straight track @ 8–10 mph – tons	2710	
On 1 in 100 @ 8–10 mph – tons	708	
On 1 in 75 @ 8–10 mph – tons	559	
On 1 in 50 @ 8–10 mph – tons	386	
On 1 in 25 @ 8–10 mph – tons	182	

Kitson & Co. Works No. 5325/1920 was the last built 0–6+6–0T for the Girardot Railway and when photographed had been renumbered Centrales 144. It had been retired by February 1954 and when photographed carried the legend "Centro de Ingenieros Militares Francisco J. Caldas"—being used for training purposes by the engineering section of the army. G. Diaz.

No. 25 had been built as a Type 1 0–6+6–0T with its cylinders at the rear of each steam bogie. It was one of two later converted to pseudo Type 3 by the substitution of a new rear bogie rotated to give cylinders at the outer ends configuration. G. Diaz.

B. 2−6+6−2T Colombian Government Railways−Girardot Railway

Dewhurst Kitson Meyer No. 24 (Kitson & Co. Works No. 5400/1927)−the first of four highly successful 2−6+6−2T.
via P. K. Dewhurst.

In 1924 the eight railways owned and operated by the Colombian Government (totalling nearly 700 miles) were brought under single control and the separate Mechanical Engineers departments combined under the leadership of P. C. Dewhurst−the Government C M E−who set about preparing a series of standard designs for locomotives, carriages and wagons. (P. C. D. had arrived in Colombia on 27 June 1923; he left the country early in 1929).

In October 1925 P. C. Dewhurst made a bid to be granted permission to order some large 2−6+6−2T Kitson Meyer locomotives which he had designed in conjunction with Mr Brocklebank (of Kitson & Co.). The following is extracted from the Colombian Government file and reproduced courtesy Gustavo Arias and P. K. Dewhurst:

Articulated Locomotives & The Girardot Railway

"With respect to the ordering of modified Kitson Meyer locomotives for service on the upper part of the F. C. de Girardot, I consider it necessary to make some observations.

In the first place I would mention that since 1910 I have been in charge of mechanical departments of mountain railways where all the special locomotive problems in the operation, reliability, cost of repairs etc. due to such severe conditions have been under my observation for practically 16 years.

This fortunate experience, which very few Chief Mechanical Engineers have had−has caused me to be considered to some extent an authority on locomotive design and practice for mountain lines; it has also enabled me to develop and design a type of locomotive (now known as the "Dewhurst" type) in which has been obtained the greatest power together with economical operation and repair that it is possible to obtain for any given gradient and curve without having recourse to an articulated locomotive.

In any mountain railway there arrives a time, depending on the severity and continuity of its curves, when the density of its traffic and the necessity of handling it with dispatch and economy, when the ordinary types of locomotive must be replaced by the "Dewhurst" and after the ultimate capacity of that type has been reached it is necessary to adopt an articulated locomotive.

In the particular case of the Girardot line, the upper section−not only on account of the severity of its curves, but particularly on account of that section consisting of 60 kilometres of *practically continuous* curves−has from its opening necessitated the use of articulated locomotives.

Of articulated locomotives there are four types viz:- the "Fairlie", the "Shay", the "Kitson Meyer" and the "Garratt". The Fairlie on account of its using the boiler as part of the main framing has been abandoned for heavy main line work, whilst the Shay is a side driven geared locomotive only suitable for slow construction work.

Of the remaining two types the Kitson Meyer is the most correct theoretically and practically for sharply curved mountain railways; the Garratt being more suited to cases of exceptionally heavy trains on ordinary lines with steep gradients but where the loading gauge will not allow a sufficiently large locomotive of the ordinary types to be constructed. The "Garratt" has its bogie pivots placed in a very undesirable position and is not a true double-bogie vehicle in a similar sense to a passenger carriage as is a Kitson Meyer; further the proportion of weight on the various axles of a "Garratt" locomotive varies as the coal and water is used up when running and this does not happen with the Kitson Meyer; apart from the foregoing the "Garratt" has the disadvantage of a bad view of the line from the cab especially when running forward−and the main steam and exhaust pipes are very close to the rails and so liable to damage; these disadvantages do not exist in the Kitson Meyer type.

As an instance of the care in detail design and excellence of workmanship in the Kitson Meyer I would instance the case of a serious derailment which ocurred in my presence this year at Anapoima, where the engine of the passenger train was derailed at the points; 10 of the 12 wheels leaving the rails and the engine tilted over considerably. I rerailed this engine in 40 minutes with the assistance of another engine, the derailed engine was rerailed by the combined force of itself and the other engine no jacks being used. On examining the rerailed engine it was found absolutely undamaged and continued with the passenger train to Facatativá gaining some minutes on the run.

Reference should be made here to a special type of locomotive the "Mallet" as this is often mistaken in Colombia for an articulated locomotive, whereas it is really only **semi-**articulated. It is not at all suitable for sharply curved mountain lines, being theoretically and practically incorrect in its functioning on curves of different radii—particularly when running backwards. As is well known now in Colombia it cannot be got to operate successfully on our lines and this could have been forcast by any locomotive engineer having the necessary experience.

The new Kitson Meyer locomotives projected are for the object of quickening the up journey from La Mesa to Facatativá; for this reason they will be about 25% larger than the existing engines, but as the proportions of the new engines will be better than the existing ones, the increase in power available for the actual haulage of trains and the general efficiency in service will be much greater than the increase in size.

It should be explained that the existing Kitson Meyer engines on the F. C. de Girardot have exceptionally small boilers, and this was pointed out by me to the builders when I made a study of the design in England just before coming to Colombia; Kitsons informed me that this was caused by the restricted axle-loads due to the 45lbs rails and Coopers E 30 bridges and I found this statement to be correct as the existing engines are the limit of weight allowable for the existing conditions. This small boiler capacity caused difficulties from the commencement and was only over-come by using an intense draught which in its turn has caused considerable repairs to be necessary and a short life to the fireboxes and tubes; but in spite of this the engines have conducted with every success and **perfect safety,** a service without equal in this or many other countries. There are also certain detail inconveniences in the existing engines, mostly in regard to ease of inspection and repairs.

All the existing shortcomings will of course be eliminated in the new design now being prepared here which in addition to the fundamental advantages of the Kitson Meyer type will also have incorporated all suitable features of the Dewhurst type and the new locomotives will have the advantages of both. Of course advantage is being taken of the new Coopers E 40 bridges and 60lbs rails with which the Girardot line will be equipped by the time the new locomotives are ready for service.

It is however impossible to design a satisfactory locomotive for the upper section of the Girardot line without using the fundamental arrangements and many of the details, of the Kitson Meyer locomotive. As an instance of one of these points may be mentioned the repression brake system fitted to all Kitson Meyer locomotives; by means of this all trains on the Girardot line are controlled down the long gradients entirely without applying the air-brake (which is used only for stopping at stations etc.) the saving in brake-shoes and gear also longer life of the wheels and other parts of all the coaches and cars is enormous, quite offsetting in a year any difference in first cost which may be occasioned by purchasing this special type; further as is also well known, no train in charge of a Kitson Meyer engine has ever run away on the Girardot line which under other conditions might have been a dangerous line to operate.

All the engineers, officials, and others on the Girardot Railway are in accord with the continuation of the Kitson Meyer locomotives on that-line.

Bogotá 21/10/25
P. C. Dewhurst".

Rear view of 2−6+6−2T No. 24 in the paint shop of Kitson & Co., Leeds.   *via P. K. Dewhurst.*

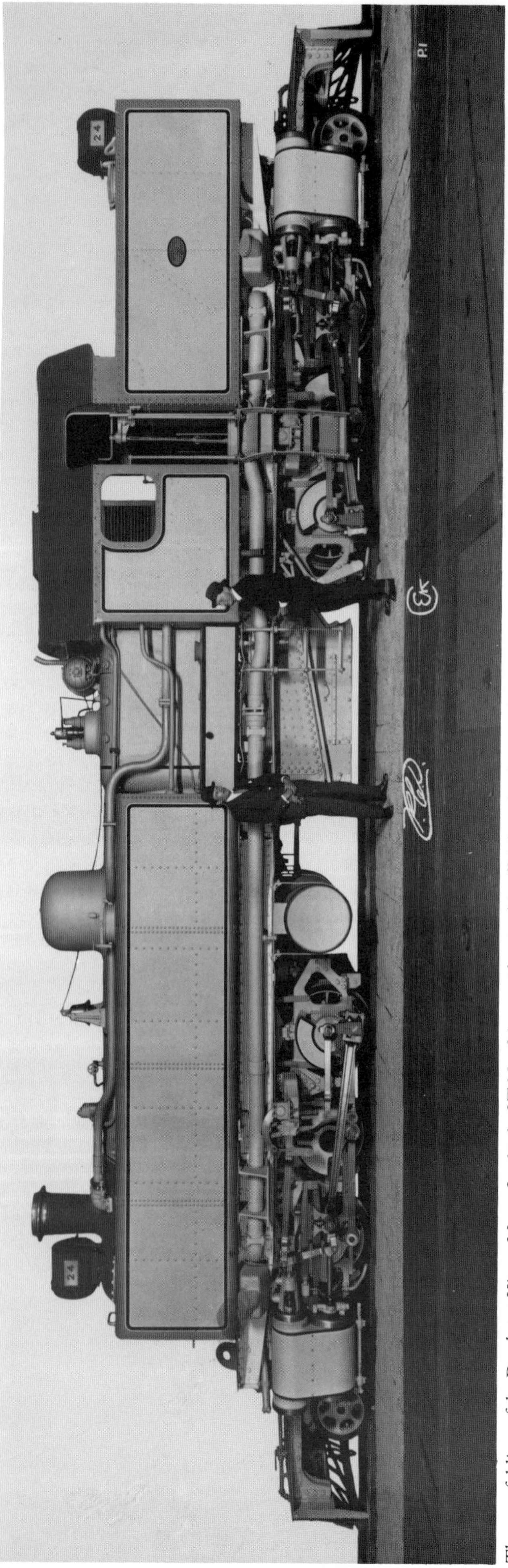

The graceful lines of the Dewhurst Kitson Meyer 2−6+6−2T No. 24 are evident in this official works photograph taken on 20 December 1926 before shipment to Girardot. Mr. P. C. Dewhurst standing on the left; Lt. Col. Edwin Kitson Clark on the right. These locomotives were painted in two shades of green: a light colour on the side tanks and bunker, and a darker shade for the rest including the boiler jacket and steam dome. Trim lines were yellow with a black outline. D. Binns collection.

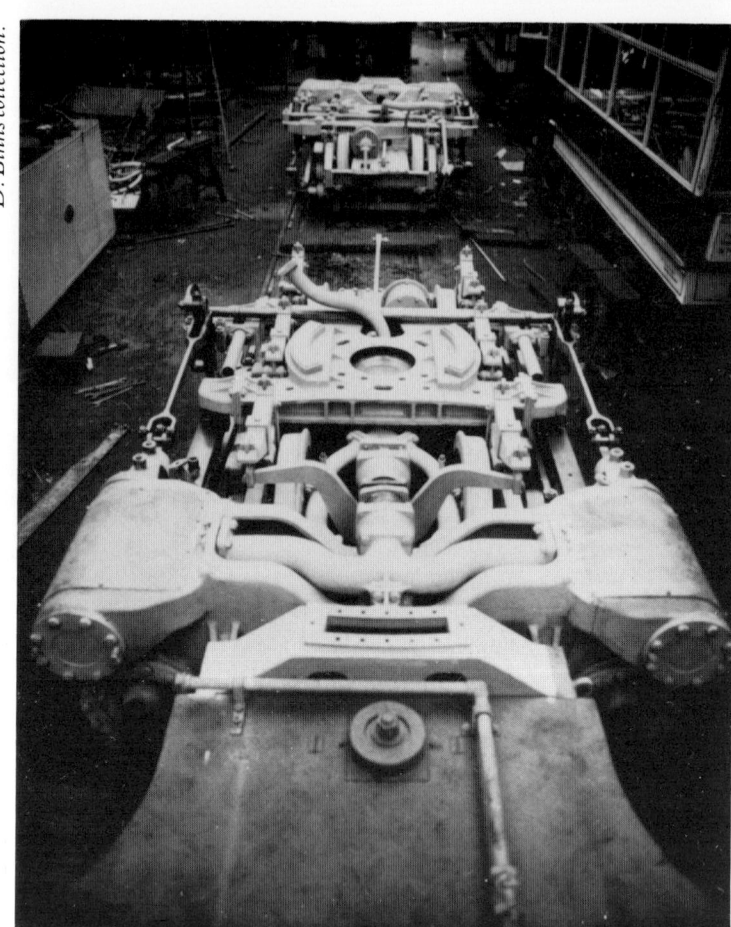

Kitson & Co. Works No. 5400−leading bogie viewed from both ends. via P. K. Dewhurst.

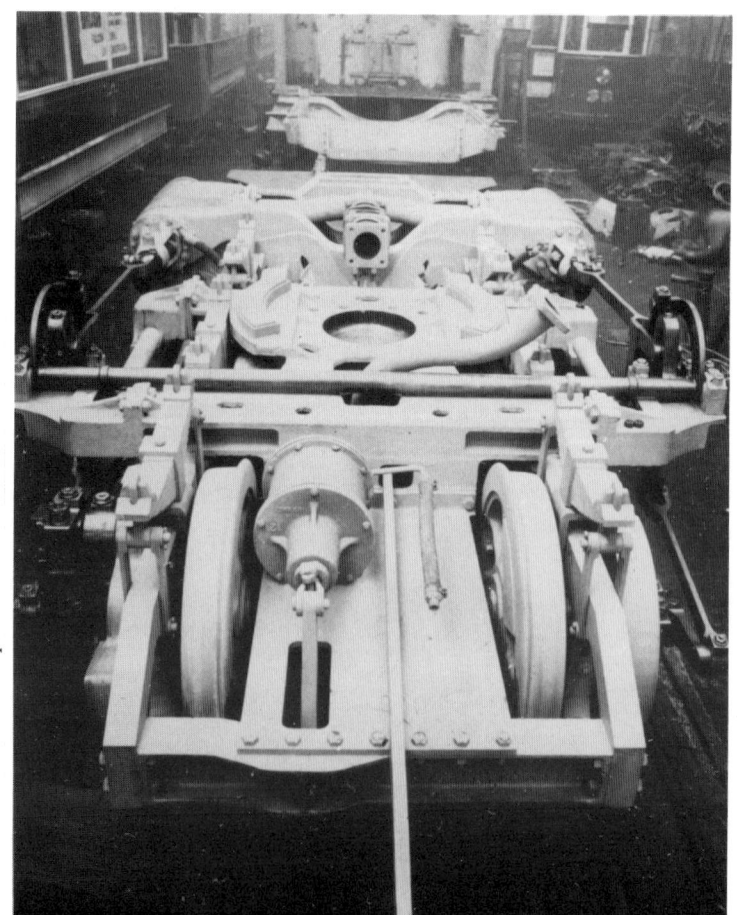

Kitson & Co. Works No. 5400—trailing bogie viewed from both ends.
via P. K. Dewhurst.

Footplate view of Kitson & Co. Works No. 5400.

via P. K. Dewhurst.

In the event Dewhurst was successful in his bid and in 1927 the superb 2−6+6−2T of type 3 were supplied for use on the hill section of the Girardot Railway. These were considerably larger and more powerful than previous locomotives, being introduced to allow heavier passenger trains to negotiate the hill section at faster speeds, and it is interesting to note that with the introduction of these locomotives, the journey time from Girardot to Facatativá was reduced by one hour. The design and specification was due to Mr P. C. Dewhurst in conjunction with Kitson & Co. also and these 2−6+6−2T were the finest Kitson Meyer locomotives ever built, ranking amongst the world's most efficient articulated power. Considerable improvements were incorporated in the design including the best of both British and American locomotive practice. Outside bar frames were used and the spring gear and brake apparatus was easily accessible for servicing. The brake gear was arranged independantly on each steam bogie to avoid any possibility of pulling the bogies onto a straight line when a brake application was made on a curve. In order to secure necessary curving properties the end pony trucks had three point suspension hangers. All the coupled wheel springs of each bogie were equalised and compensating levers connecting to the pony trucks were provided. A divided side tank was employed, the rear tank containing 1050 gallons and the side ones 1450 gallons. A modified Belpaire boiler was fitted with the crown of the inside firebox and the roof of the outer shell plate curved—a neccessity on locomotives working over lines with super-elevation of the rails in order to obviate scorching of the firebox crown. The inside firebox, tubes and flues were mild steel as were the roof stays with cup heads inside the firebox. The water space stays were of best Yorkshire iron. The tubes had their beads electrically welded to the tube plate

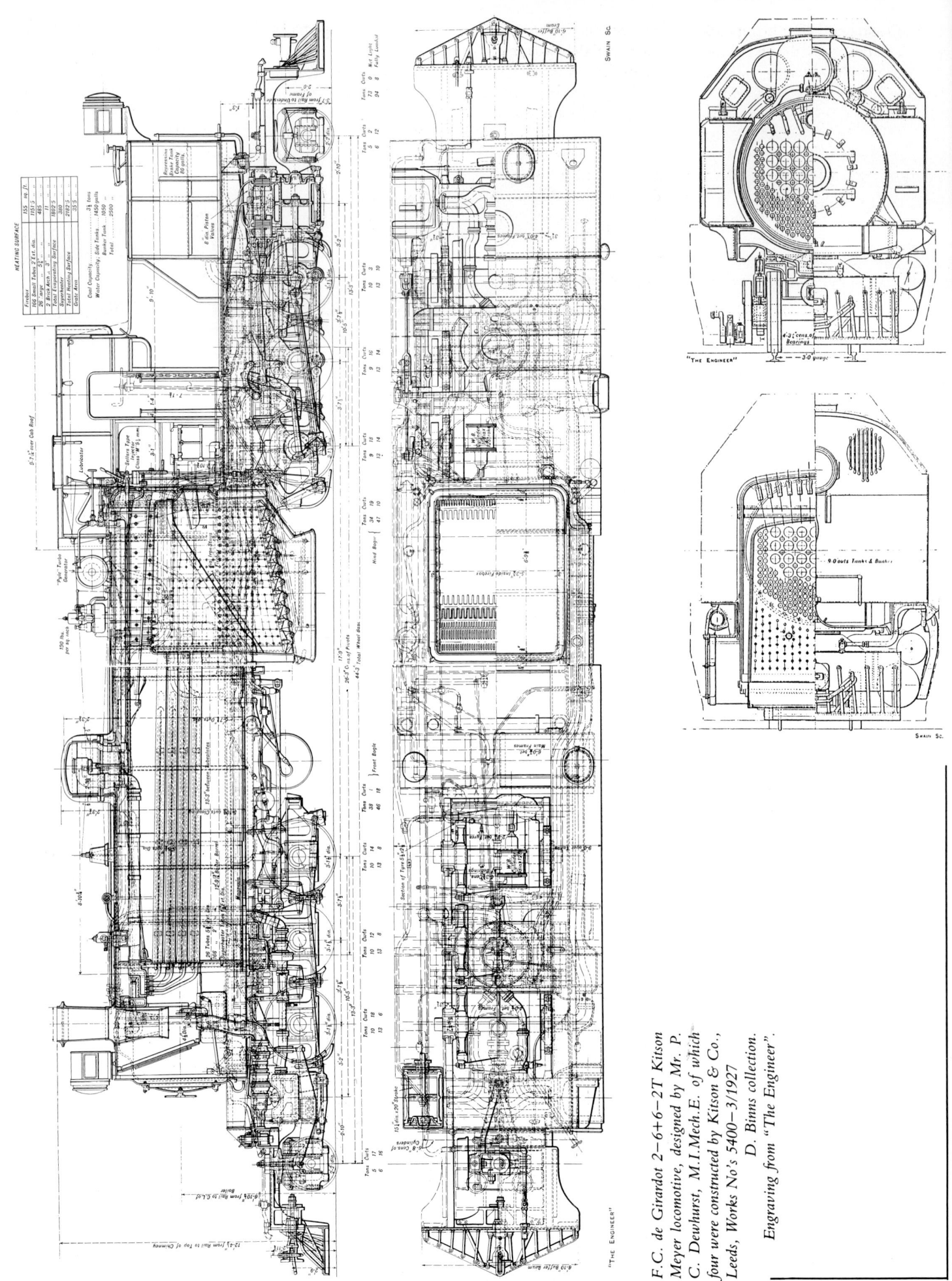

F.C. de Girardot 2–6+6–2T Kitson
Meyer locomotive, designed by Mr. P.
C. Dewhurst, M.I.Mech.E. of which
four were constructed by Kitson & Co.,
Leeds, Works No's 5400–3/1927
        D. Binns collection.
Engraving from "The Engineer".

at the firebox end after being expanded and beaded in the usual way. The firebox had two arch tubes which partially supported the brick arch and also assisted in increasing circulation of water in the boiler. The centre pair of driving wheels on each steam bogie were flangeless but 6½in wide whilst the others were 5½in. The Colombian Government requirements were comprehensive and included a 2½kw capacity electric generator to supply power for headlight, cab lights and a 10 coach train. Electric lights were also provided beneath the water tanks and over the motion on both side of the locomotive. Other fittings included a Webb fire hole with double sliding doors, rocking finger grates and a removeable smokebox front with small smokebox door secured by dogs. There was a steam turret over the firebox in the cab, pop safety valves mounted 6in above the boiler barrel, spark arrester, ash ejector, injector top feed delivery, extra cleaning holes near the front and back tube plates, a mud door beneath the barrel and Laird type cross-heads and slide-bars. The regulator was of the double beat poppet type with pull out handle which permitted fine adjustment in steam admission—a feature neccessary on railways with frequently differing grades and curvature. A repression brake was fitted. These locomotives represented a considerable advance over the previous Kitson Meyer's then in service and were indeed powerful machines for so narrow a gauge with a tractive effort at 85% boiler pressure of 40,000lbs. The weight in working order was 94 tons 8 cwts.

KITSON MEYER TYPE 2−6+6−2T SUPPLIED TO THE COLOMBIAN NATIONAL RAILWAYS

Kitson No.	Date Built	Steam Trials	No. as Built	1927 to 1929	1930 to 1931	1931 to 1937	1938 to 1945	1946 to 1953	1953 on	Notes
5400	1927	14/12/26	24	24	24	24	31	35	154	A, B
5401	1927	6/ 1/27	25	27	27	27	34	34	152	A
5402	1927	21/ 1/27	26	25	25	25	32	32	151	A
5403	1927	7/ 2/27	27	26	26	26	33	33	153	A

Note: A. KMs 5400 thru 5403 were built at Kitsons as No's 24 through 27 respectively but were re-erected at Girardot as No's 24, 27, 25 and 26 respectively.

Note B. In 1946, C/N 5400 appears with No. 35 and the story, from retired Girardot locomotive engineer C. Romero T. is that on Good Thursday, 1940, #31 derailed between Tocaima and Pubenza with a balance of 20 people dead and many injured: No. 31 had had three other accidents and the crews had the locomotive renumbered because they attributed the accidents to bad luck brought about by her number (31 is 13 inverted, you see)! So she became Girardot 35. The group was renumbered Centrales 154 (C/N 5400), 152 (5401), 151 (5402) and 153 (5403), when the Centrales Division of the Colombian National was formed in 1953.
All appear with C/N and previous No's on Roster February 1954 Division Centrales (Girardot, Cundinamarca, Norte 2, Nordeste, Sur, Tolima. Not included Dorada and Ambalema).

No. 35 was the original Kitson Meyer 2−6+6−2T supplied to the F.C. de Girardot. It was photographed in the period 1946-1953 whilst numbered 35 and despite some 20 years hard work looked extremely smart. These Dewhurst designed locomotives were handsome in appearance and the most successful of the Kitson Meyer types.

G. Diaz.

Kitson 5400/1/2/3 built as type 3, with 4 cylinders 15¼×20, 3ft 1½in driving wheels, 2ft 2in carrying wheels, rigid wheel base 7ft 3in, total wheel base 44ft 3in, boiler pressure 190lb, boiler diameter 5ft 7¾in, firebox length 7ft 6in width 5ft 11in, tubes 166×2in×13ft 4½in and 26×5⅜in.

This commercial post card shows the first ordinary passenger train to cross the newly completed bridge over the River Magdalena at Girardot at 10-54am on 1 January 1930. The train comprised 14 carriages hauled by Kitson Meyer 2−6+6−2T No. 26 (Kitson & Co. Works No. 5403/1927). The bridge, built by Sir W. G. Armstrong Whitworth & Co. Ltd., is too long for the picture—in reality nearly ½km with the main span 124 metres long.

Fotografia Saray Hnos, Colombia, via C. J. Walker.

Kitson Meyer 2−6+6−2T Centrales No. 152 was photographed at Nordeste shop circa 1960−another 2−6+6−2T is in the right background. Note the modified sandboxes when compared with the picture of No. 35 on page 57.

G. Diaz.

C. Kitson Meyer 2−8+8−2T Type 3 Supplied to the Colombian National Railway

The final two Kitson Meyer locomotives delivered to the 3ft 0in gauge Girardot section of the Colombian National Railways were a pair of 2−8+8−2T. The order for these was placed with Kitson & Co. but as they were unable to undertake construction and had already disantled the heavy cranes that would be required to lift these locomotives, it was passed to Robert Stephenson & Hawthorn of Darlington. All the parts were made in Leeds and these locomotives did in fact carry Kitson & Co. Works No's 5471/2 of 1935. They were an enlargement of the 1927 Girardot 2−6+6−2T with a considerable increase in power and at 85% of the 205lb boiler pressure, the tractive effort was a formidable 58,564lbs—a real power house on so narrow a gauge. In real terms they could haul 330 tons up 1 in 22 at 10 mph The total weight in working order was 130 tons 8 cwt and the maximum axle load only 14.55 tons, a slight increase over the 13.75 tons of the 1927 2−6+6−2T. An interesting feature was the return to a full length side tank which contained 2500 gallons of water, a

further 1500 gallons being accommodated in the rear bunker tank, along with 8 tons of coal or 1100 gallons of fuel oil. With an overall height of 12ft 4½in, a width of 9ft 0in and a length of 66ft 4¾in riding on 3ft 0in gauge track, these were locomotives to be reckoned with and indeed were considerably more powerful than any British 4ft 8½in gauge locomotive (the LNER Beyer Garratt excepted). All the standard Colombian Government features were incorporated into this design including bar frames which left the space between the outside frames almost clear with the brake and spring gear easily accessible for adjustment. A Pyle National electric generator supplied power for headlights, cab and train lighting and equipment included Alco type K air operated reverse gear, an 8½in cross compound air compressor, 3−2½in Crosby safety valves and a repression brake. The wheels of the second coupled axle in each unit had thin flanges whilst the wheels of the third coupled axle were flangeless to enable them to negotiate 213ft minimum radius curves.

GIRARDOT RAILWAY 1938 ROSTER−OFFICIAL

Kitson Works No.	R.S.H. No.	Year	Original No.	1935 to 1937	1938 to 1945	1946 to 1953	1953 on	Note
5471	4110	1935	56	56	73	83	172	A
5472	4111	1935	57	57	74	84	173	A

Note: A. Shown with C/N and previous numbers on Roster February 1954. Division Centrales. Both then oil fired. Withdrawn in the period 1958-60.

Kitson Meyer 2−8+8−2T No. 57 was Kitson & Co. Works No. 5472 (RSH 4111/1935) and this front ¾ view shows the extremely handsome lines of these exceptionally powerful 3ft 0in gauge locomotives.
D. Binns collection.

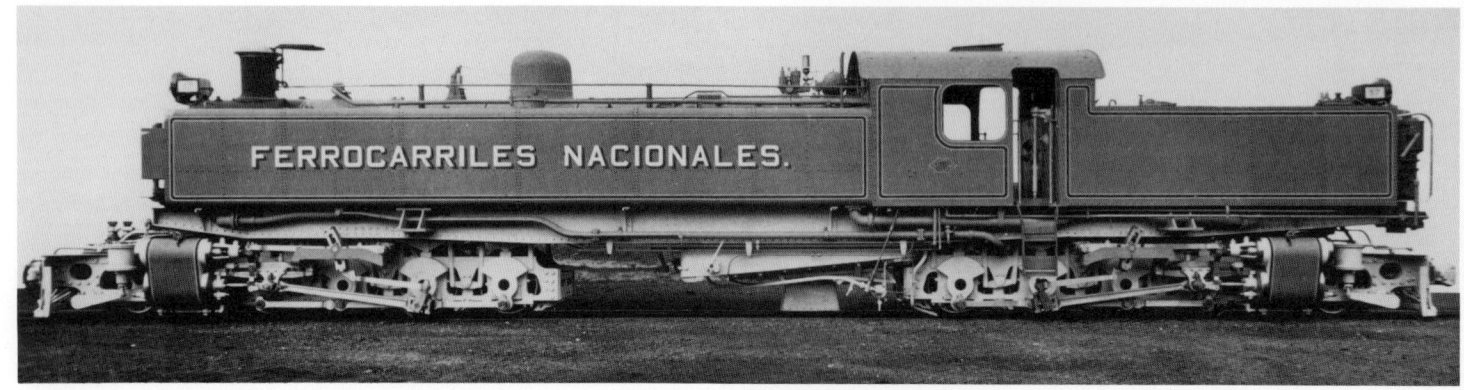

FERROCARRILES NACIONALES.

Broadside view of 2–8+8–2T No. 57. The two locomotives of this type never performed as hoped on the Girardot and were transferred to other duties where the curves were less severe. The one serious problem was that their 8-coupled wheelbase was just too much for the severe Girardot curves. P. C. Dewhurst noted in October 1925: "the upper section—not only on account of the severity of its curves, but particularly on account of that section consisting of 60km of **particularly continuous curves**—*has from . . ." In 'The Locomotive' for 15 February 1927 he states "maximum gradient of 1 in 25 is continuous for considerable distances coupled with almost continuous curves of 260ft radius". A USA Department of Commerce survey made in 1925 gives the following on the F.C. Girardot: "Curves—The minimum curve radius at km 85 (between Hospicio and Anolaima) is 65 metres (213ft); this curve is uncompensated, and there is almost no tangent between it and the next curve. The sharpest curves elsewhere have a radius of 70 metres (230 ft) but none of them are compensated". The Girardot presented a really formidable set of difficulties and PCD's 2–6+6–2T were about as large as practicable in the circumstances.* *D. Binns collection.*

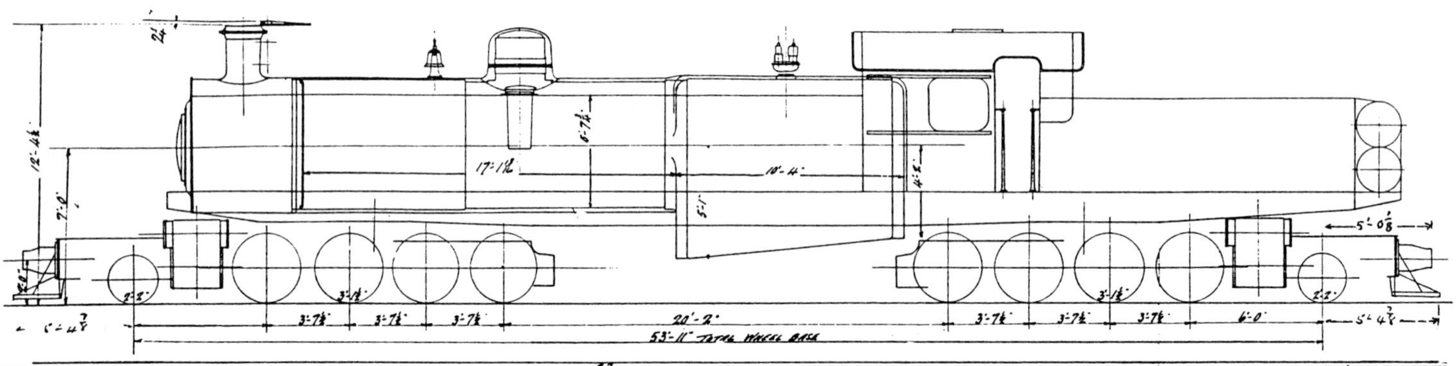
RAILWAY CO. GIRARDOT. GAUGE OF RAILS 3-0 ENGINE No. 157

DIAMETER OF CYLINDER 17¾" LENGTH OF STROKE 20"
Distance of Centres 6'-10" Length of Con. Rod

WEIGHTS OF ENGINE (FRONT BOGIE	Full Tons Cwt Qrs	Empty Tons Cwt Qrs
On Leading Wheels,	9 10 0	8 6 0
„ „ „ 15 CWTS ADDED	14 16 0	12 6 0
„ Driving „ TO FRONT BOGIE.	13 19 0	10 17 0
	13 10 0	10 14 0
„ Trailing „	12 15 0	9 17 0
TOTAL WEIGHT	64 10 0	52 0 0
Coal in Firebox,		
„ Bunker, C. ft		
Water in Boiler,		
„ Tank, Gallons,		
Sand, C. ft.		
TOTAL WEIGHT ON BOTH BOGIES	130 5 0	100 13 0
WEIGHTS OF ~~TENDER~~ HIND BOGIE	13 1 0	8 15 0
On Front Wheels,	14 1 0	9 19 0
„ Middle „	14 3 0	10 7 0
„ Hind, „	14 11 0	11 12 0
	9 19 0	8 0 0
TOTAL WEIGHT	65 15 0	48 13 0
Water in Tank, Gallons,		
Coals, C. ft.		
TOTAL WEIGHT		

TUBES (STEEL. 5⅜" EXT. 2") No. 30 176.
Ex. Diameter at Firebox 4½" 1 13/16" at Smokebox 5¼" 2¼"
Thickness „ 9 W.G. 11 W.G. „ 9 W.G. 11 W.G.
Distance of Centres „ 6¼" 2·7" „ 6¼" 2·7"
Length between Tube Plates, 17'-7"
Heating S. in Tubes, Sq. ft 1620 742. Firebox Sq. 205
Total „ 2567 Grate „ 51.
+ ELEMENTS 640
Working Pressure, 205 lbs per Sq. in.

THICKNESS OF PLATES
Boiler Barrel 23/32" Smokebox Tube Plate ¾"
Firebox casing Sides 9/16" top 9/16" back 9/16" front 9/16"
Firebox (STEEL) Tubeplate ⅝" top ⅜" back ⅜" sides ⅜"
Engine Frameplate (3½" BOGIE) (⅞" MAIN.)
„ Tanksides 3/16" top 3/16" ends 3/16" bottom 5/16"
AXLE journals dia. 7½" length 7" material
„ „ BOGIES 5" „ 9"
TIRES thickness 2½" 2¼" breadth 6¼" DRIVERS. 5¼" COUPLED.
„ „ BOGIES 2½" „ 5½" „
INJECTORS size „
PUMPS how worked „
Tender Frameplate outside inside
„ Tanksides top ends bottom
„ journals dia. length
„ TIRES thickness breadth

DETAIL WEIGHTS OF ENGINE — Tons Cwts Qrs Lbs
Leading Wheels and Axles,
Driving „ „
Trailing „ „
One Leading Axlebox,
„ Driving „
„ Trailing „
„ Connecting Rod
Coupling Rods on One Side,
One Leading Spring,
„ Driving „
„ Trailing „
Boiler with Firebox and Tubes,
Frame and Cylinders,
Chimney,
Eccentric Pulleys (Complete),
„ Straps „
„ Rods „
Reversing and Lifting Links,
One Piston and Rod
„ Crosshead and Slide Blocks,
Cast Iron Footplate,
One Pair Cyl. Covers and Glands,
„ Bogie with Wheels and Slides,

DETAIL WEIGHTS OF ENGINE Contd — Tons Cwts Qrs Lbs
Tanks,
Bunkers,
One Set Brake Gear (Complete)
„ „ Spring „ „
Cab Cwts.,
Splashers,

DETAIL WEIGHTS OF TENDER
Frame,
Tanks,
One Sets Wheels and Axles,
„ „ Spring Gear,
„ „ Brake „
„ Spring,
„ Axlebox,

Gustavo Arias recalls "I still remember them in their green livery and when I was young I considered them the most beautiful locomotives ever built. But their operation, sad to say, was really a disaster. Perhaps the track had deteriorated when they were put in operation, but the truth is that they were too heavy and rigid (both horizontally and vertically) and derailed frequently. The crews did not like them and used to say that the "Stephensons" were the best rail straighteners ever devised. They were also retired before 1960. (When I joined the CNR as an engineer in 1959 they were already retired at the Nordeste Shop, in Bogotá)."

During 1955/6 there was an exchange of correspondence between P.C.D. and Douglas S. Purdom (Consulting Engineer) concerning the latter's visit to Colombia in connection with the two 2–8+8–2T. *Extract from D. S. Purdom's letter dated 20th December 1955.*

. . . Last August I made a trip to Colombia on behalf of R. Stephenson and Hawthorns, for whom I act as a technical representative here, and remembering your connections with the railways of that country, I thought you might be interested in hearing something of my visit. . .

My mission was concerned with the two large Kitson Meyer locomotives built in 1935 which had recently been fitted with Laidlaw-Drew oil burning systems. These were not giving satisfaction so I was asked to investigate.

Having only ten days or so I was not able to make exhaustive trials or extensive modifications, but gradually the trials I did make improved and on the last occasion we took 210 tons up the very steep bank from La Mesa with reasonable success. At this point I had to leave after indicating how matters would be still further improved. The line from Girardot up to Bogotá is certainly a stiff test for any kind of motive power.

This was my first experience of the Laidlaw-Drew system and I must confess I was not terribly impressed. The two vertical burners gave a nice short saucer shaped flame and the effect looked like a good coal fire, but they are of the internal mixing type with a very narrow orifice, and I found always a definite tendency to clogging after a couple of hours or so of hard working.

I still have not seen anything to beat a properly designed Mexican trough installation, especially under rough conditions and poor maintenance as in Colombia.

The shops and shed at Girardot are in an appalling condition. New buildings are under construction across the river at the old wagon shop, but they will not be ready for a long time.

At Facatativá and Bogotá conditions are not much better. It is the usual story with State owned S. American railways—no money for anything and a top heavy burocracy muddling along. A fairly good service of small diesel railcars is maintained between Bogotá, Girardot and other points. . ."

Extract from PCD's reply to D. S. Purdom's letter dated 20th December 1955.

"Now to your letter: The very full account you send me of loco matters in Colombia has interested me greatly (almost nostalgia) and I feel a little proud that I am still remembered there and that my locos are still behaving themselves. My time in Colombia was, as you will know from having visited some of the lines and seen their various loco operating problems, a most enjoyable and instructive experience designing for such lines. Especially at the time when nearly all of the lines were being extended or newly made altogether. Two of my very distant lines, although I designed locos for them, I never managed to visit during my five years in Colombia.

I see that you had a go at the Laidlaw-Drew oil burning equipment on the Girardot 2−8−0+0−8−2 Kitson Meyer—Dewhurst engines by R. Stephenson: I note 210 tons from La Mesa to Facatativá. I note your summing up of the Laidlaw-Drew equipment and I certainly agree with your views. In fact I think it hopeless—especially when for locos we have much better arrangements available. There is, in my opinion (and I burnt oil in Colombia on one or two of the lines and went to 100% oil in Uruguay) nothing to equal the burner at the front of the firebox throwing the flame towards the rear.

As you say, workshops get little attention in places like Colombia and when I say that the *commencement* of the workshops at Flandes for locos was in my time, we have good proof! . . . . . . etc etc."

Extract from D. S. Purdom's reply dated 3rd August 1956.
". . . The two Kitson Meyer—Dewhurst engines I thought were a magnificent job and they ride like Pullman cars. They must surely be about the maximum limit of power that can be put on the 3'-0" gauge.

Now that I come to think of it, I believe Robayo told me that the metre lines were being converted to 3'-0" gauge, which at the time struck me as being rather a retrograde step, though no doubt taken in the interests of standardization. (Dr. A. Robayo was then Chief of Traction).

The sequel has been that, in spite of Robayo carrying out faithfully all the suggestions I left with him for improving the

working of the Laidlaw-Drew system, it was found impossible to maintain sufficient steam with them so they have been discarded and replaced by trough burners in which form I feel sure the engines are doing an excellent job."

In 1935 a simple Mallet 2−8−8−2T (Girardot No's 58, 82, 72, 171) was purchased from Baldwin as a comparison against the Kitson 2−8+8−2T but no records are available for comparison.

In reference to my original book (published by Wyvern Publications), Gustavo Arias writes:
"I must mention that the color painting of Girardot 83 by W. Meadway is very good. However, and of course, the Artist had no way of knowing it, no Colombian locomotive had the underframe painted red (only the pilot beam, and in some cases, the side of the running board). Also when the units were painted black, the lettering was yellow without a shadow, and may have read: Ferrocarriles Nacionales , Girardot (underneath)."

Railway	Colombian National
Gauge	3ft. 0in.
Wheels	2-8+8-2T
Maker	Kitson/R S & H
Works No.	K5471/2
Year	1935
Cylinder position	outer
Cylinders – inches	17¾ × 20
Boiler Pressure – lbs./sq. in.	205
Heating Surface	
Firebox – sq. ft.	205
Tubes – sq. ft.	2362
Total – sq. ft.	2567
Superheater – sq. ft.	640
Total H S – sq. ft.	3207
Grate area – sq. ft.	51
Driving wheel – diameter	3ft. 1½in.
Other wheels	2ft. 2in.
Rigid wheelbase – first group	10ft. 10½in.
Rigid wheelbase – second group	10ft. 10½in.
Total wheelbase	53ft. 11in.
Water – galls.	4000
Coal – tons	8 or 1100 galls oil
Weight empty – tons cwt.	99 12
Weight in working order – tons cwt	130 4¾
Weight adhesive – tons cwt.	
Tractive effort – lbs.	51,600 (75%) 58,564 (85%)
Overall – height	12ft. 4½in.
,, – width	9ft. 0in.
,, – length	66ft. 4¾in.
Hauling capacity on straight track @ 8–10mph – tons	
On 1 in 100 @ 8–10mph – tons	
On 1 in 75 @ 8–10mph – tons	
On 1 in 50 @ 8–10mph – tons	
On 1 in 25 @ 8–10mph – tons	
Boiler diameter	5ft 7in
Firebox length	10ft 4in
Firebox width	5ft 11in
Tubes	176 × 2in × 17ft 9in 30 × 5⅜in
Weight leading axle	9tons 10cwt
Weight trailing axle	9tons 19cwt
Weight on driving wheels	110tons 15¾cwt

ROUTE FROM THE PACIFIC TO BOGOTA

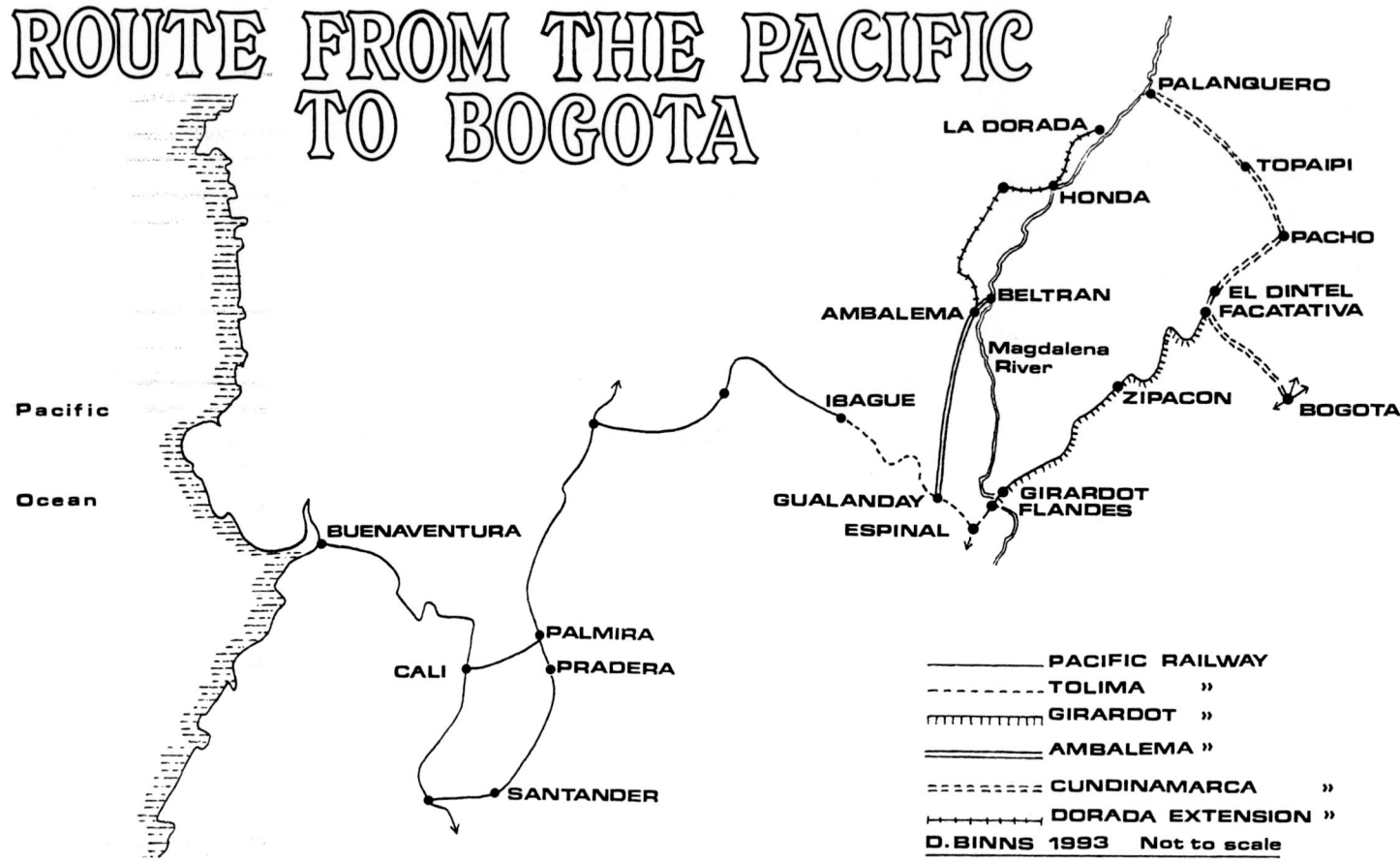

During the mid 1920s a trunk line was being developed connecting the Pacific and Atlantic coasts with the capital city of Bogotá. This trunk line had as its nucleus the 3ft 0in gauge Pacific Railway (Ferrocarril del Pacifico) system and by means of this line from the Pacific coast port of Buenaventura to Ibague, thence over the 3ft 0in gauge Tolima Railway from Ibague to Flandes, then over the 3ft 0in gauge Girardot Railway from Flandes to Facatativá (these 3 railways being Government owned), and finally via the Cundinamarca Railway from Facatativá to Bogota (a line which was owned in part by the Department of Cundinamarca and the Government).

Previous to 1930 it was necessary to change trains at the Magdalena River (near Girardot) because the bridge there was still under contruction—on completion the Tolima Railway would become connected to the Girardot Railway.

Construction of the metre gauge Ferrocarril de la Sabana had started in 1882 and its line from Bogotá to Facatativá (55km) opened in 1889. On 31 December 1921 an agreement was entered into between the Department of Cundinamarca and the Colombian Government whereby the Ferrocarril de Cundinamarca was organised, its purpose to take-over and to jointly operate the F.C. de la Sabana, and its short 8km extension from Facatativá to El Dintel. A new line was to be constructed from El Dintel to the Magdalena River at Palanquero, near La Dorada.

In order to obviate the gauge change at Facatativá the F.C. de Cundinamarca converted its line between Bogota and Facatativa from 1 metre to 3ft 0in gauge in 1924 so allowing passengers and freight from Girardot to reach the capital without changing trains. The extension to the Magdalena River was at the time

started in 1 metre gauge (the gauge of the Sabana at the time), i.e. the section from Facatativá to El Dintel. When the Sabana was narrowed, the upper part of the Cundinamarca to Palanquero was continued in 3ft 0in gauge. The lower metre gauge section between El Dintel and Facatativá was subsequently narrowed and the line reached Puerto Salgar (on the Magdalena) in March 1936. The maximum grade between Facatativá and Bogotá was 1 in 50 with minimum radius curves of 200 metres. Between Facatativá and El Dintel the maximum grade was 1 in 33. Wood sleepers(later steel), dirt ballast and rails varying between 45 and 75lbs/yd were in use. "Railways of South America"—Part II notes that "At Bogota there is a moveable semaphore signal; otherwise coloured flags and lanterns are used". At the end of 1925 the Company had 20 locomotives in use including 2 new Mallet semi-articulateds.

During 1928/9 Kitson & Company constructed two different types of 2−6+6−2T Kitson Meyer for use on the upper section of the Cundinamarca Railway, presumably the then uncompleted extension from El Dintel to Palanquero—both to work on 3ft 0in gauge track laid with 198ft radius curves. Both types were designed with view to easy conversion to 1 metre gauge. The 1928 delivery to the Cundinamarca Railway comprised two "baby" Kitson Meyer locomotives designed for use on new and unconsolidated track. These were built to Kitson & Company designs based on the earlier 0−6+6−0T supplied to the Girardot Railway but modified by P. C. Dewhurst to include leading and trailing pony trucks plus the comprehensive requirements laid down as standard practice by the Colombian Government Railways. As in the 1927 Girardot heavy 2−6+6−2T, the CR locomotives had a 2½kw electric generator and the brake gear of each steam bogie was self contained. These two "baby" Kitson Meyer's were of course type 3 locomotives with the cylinders at the outer ends of each unit. Perhaps the term "baby" Kitson Meyer is mis-leading for these were indeed powerful locomotives producing 27,614lbs tractive effort at 85% boiler pressure and weighing 70 tons 4 cwt (in working order) were heavier than the 0−6+6−0T on which the design was based. Only one of the small 2−6+6−2T was delivered, the other being "lost in transit". The larger Cundinamarca Railway Kitson Meyer 2−6+6−2T was built as a replacement for the "lost" locomotive but was much larger being almost identical with those supplied in 1927 to the Girardot Railway— the principle difference being the cab side sheets which were of increased height due to the more generous loading gauge of the Cundinamarca section. General dimensions and details were as the 1927 locomotives but with a slight increase in working order weight to 96 tons.

KITSON MEYER TYPE 2−6+6−2T THREE SUPPLIED TO THE CUNDINAMARCA RAILWAY

Kitson Works No.	Year	Steam Trials	Original No.		Notes
5416	1928	4/ 9/1928	13	Renumbered 11	A
5417	1928	13/ 9/1928	11	Lost in transit	
5431	1929	15/10/1929	14	Renumbered 12, then 212 and finally Centrales 155	B

Note: A. Shows in Chart February 1954 as retired. No. 11 but no C/N, built 1929.

Note: B Shown on February 1954 roster as no 12 (5431) retired−then coal fired. Shown as built 1931− incorrect, but may have entered service in that year. C/N shown with road No. 12. Renumbered 212 and Centrales 155. Retired circa. 1957

As construction proceeded on the Facatativá−Palanquero line the Cundinamarca leased (or bought) some (4?) of the 0−6+6−0 from Girardot (probably some of those missing in the 1946 Girardot roster). An 0−6+6−0 reported from 1909 and numbered Cundinamarca 14, appears in the locomotive chart for Centrales in 1954 as retired. This unit could only have been one of the three Girardot Kitsons of that year−either No's 4671 or 4673, but probably not Works No. 4672.

The Cundinamarca Railway had eleven locomotives in 1932 "owned by the Company and three belonging to the F.C. de

Kitson & Company, Works No. 5431/1929 was the last of three 2−6+6−2T supplied to the Cundinamarca Railway and considerably larger than its two predecessors−one of which had been lost at sea. It was photographed at Kitson's works in Leeds before painting. *via P. K. Dewhurst.*

Cundinamarca Railway 2–6+6–2T No. 13 was Kitson & Co. Works No. 5416/1928, the design of this "baby" Kitson Meyer was based on the small Girardot 0–6+6–0T. This locomotive was later renumbered 11.
D. Binns collection.

One of the "baby" 2–6+6–2T was lost at sea and replaced by Kitson & Co. Works No. 5431 of 1929 which carried the running number 14 when new. On arrival in Colombia, the Cundinamarca Railway renumbered it 12. This locomotive was virtually identical in dimensions to the three Dewhurst 2–6+6–2T used on the Girardot Railway, but had different cab side sheets allowed for by the more generous Cundinamarca loading gauge.
D. Binns collection.

Girardot". These would be Kitson Meyer 0−6+6−0Ts on loan−few details are known of these temporary borrowings but Works No. 5322/1920 (No. 18 in Girardot stock) was lent to the F.C. Cundinamarca and later was returned numbered 19. No. 5323/1920 also loaned went as No. 19 but came back as No. 18.

Railway	Colombian National	Colombian National	Colombian National
Gauge	3ft. 0in.	3ft. 0in.	3ft. 0in.
Wheels	2-6+6-2T	2-6+6-2T	2-6+6-2T
Maker	Kitson	Kitson	Kitson
Works No.	5400−3	5416/7	5431
Year	1927	1928	1929
Cylinder position	outer	outer	outer
Cylinders – inches	$15\frac{1}{4} \times 20$	14×18	$15\frac{1}{4} \times 20$
Boiler Pressure – lbs./sq.in.	190	160	190
Heating Surface			
Firebox – sq.ft.	155	106.5	155
Tubes – sq.ft.	1647.5	770.2	1647.5
Total – sq.ft.	1802.5	876.7	1802.5
Superheater – sq.ft.	380	160.5	380
Total HS – sq.ft.	2182.5	1037.2	2182.5
Grate area – sq.ft.	35.5	27	35.5
Driving wheel – diameter	3ft. $1\frac{1}{2}$in.	2ft. $10\frac{3}{4}$in.	3ft. $1\frac{1}{2}$in.
Other wheels	2ft. 2in.	2ft. 2in.	2ft. 2in.
Rigid wheelbase – first group	7ft. 3in.	6ft. $2\frac{1}{2}$in.	7ft. 3in.
Rigid wheelbase – second group	7ft. 3in.	6ft. $2\frac{1}{2}$in.	7ft. 3in.
Total wheelbase	44ft. 3in.	40ft. 4in.	44ft. 3in.
Water – galls.	2500	1753	2500
Coal – tons	3 18	$2\frac{1}{2}$	$3\frac{1}{2}$
Weight empty – tons cwt.	73 0		
Weight in working order – tons cwt	94 8	70 4	96 0
Weight adhesive – tons cwt.	81 0		82 $2\frac{1}{2}$
Tractive effort – lbs.	35.294 (75%)		35,294 (75%)
	40,062 (85%)	27,614 (85%)	40,000 (85%)
Overall – height	12ft. $4\frac{1}{2}$in.		
,, – width			
,, – length	52ft. $9\frac{1}{2}$in.		
Weight−front bogie−tons cwt.	6.16		
Weight−rear bogie−tons cwt.	6.12		
Weight−driving wheels−tons cwt.	81.0		

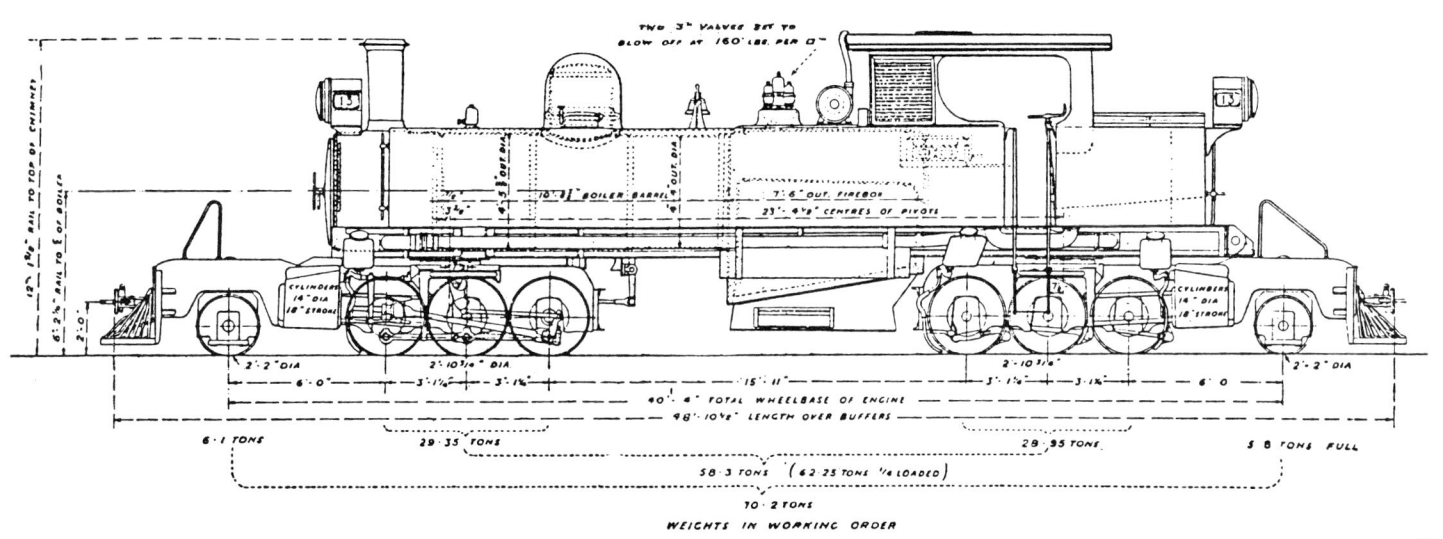

65

F.C. de Cundinamarca "baby" 2−6+6−2T Kitson Meyer No. 11 had been built originally as No. 13 and carried Kitson & Co. Works No. **5416/** *1928.*

G. Diaz.

Cundinamarca Railway large 2−6+6−2T No. 14 Works No. 5431/1929−was photographed in the works of Kitson & Co., Leeds.

via P. K. Dewhurst.

William Wheelwright had first considered the possibility of a transandine railway linking the Atlantic and Pacific coast ports in 1854, but it was to be the brothers Mateo and Juan Clark who revived the idea, drawing up plans in 1870. During the mid-1870s they were engaged on a survey for the first international telegraph line to cross the Andes, and the Clark brothers soon obtained a concession prior to making a closer inspection to find a suitable route for a railway line. Two proposals were put forward but that which eventually came to fruition ran from Santa Rosa de los Andes (in Chile), via Juncal, passing through a summit tunnel into Argentina, via the Uspallata Pass en-route to Mendoza. In 1870 there was no intention to connect the Transandine railways (Chilean and Argentinian) with Buenos Aires, the Andes crossing being solely to connect the rich Mendoza valley with Valparaiso and so bring about a considerable increase in local trade. Plans for the Chilean section were approved by the Government in 1875. Finance was to prove a problem and the law authorising the work was replaced by another in 1887. In 1888 a new body was formed—Clarks Transandine Railway Company and work finally started on the Chilean section in Los Andes on 5 April 1889. After some 30 miles had been graded work ceased for some ten years due to revolutionary problems in Chile. Eventually Clarks although much in debt, laid some 17 miles of track to Salto del Soldado.

Meanwhile Juan Clark had obtained a concession to build a metre gauge line from Buenos Aires to San Juan, via Mendoza, and (in 1878) another to extend it to the Chilean frontier via Uspallata. The Buenos Aires & Pacific Railway Company had been formed in London in 1882 and acquired the concession from the Clarks for construction of 651 miles of line from Buenos Aires to Mendoza. This was built to 5ft 6in gauge. It is recorded that the first locomotive from Buenos Aires arrived in Mendoza in 1887 concurrently with the start of construction of the Argentine Transandine section. By 22 February 1891 57 miles of line was open to traffic between Mendoza and Uspallata. Between 1890 and 1899 a financial crisis halted work and it was not until the mid-1900s that 108 miles from Mendoza, was rail connected.

Little was being achieved on the Chilean section and in 1896 the Transandine Construction Company was formed in London to purchase the section already built and to obtain from the Chilean Government a concession allowing completion of the railway to the Argentine frontier. The partners included the Bank of London, Mexico & South America who made an offer to Clark & Co. to subscribe 75% of the capital required, if the rest could be obtained elsewhere. M. Grace a New York businessman, expressed interest on condition that his Company be allowed representation in Chile—Mateo Clark agreeing the above and in due course the existing railway was

purchased and in 1903 work re-commenced. Meanwhile the Bank had pulled out leaving Grace and Clark & Co. to it. Half the necessary capital was then subscribed by the London firm of Morgan.

Work recommenced in 1903 after 12 years of inactivity. The Chilean Government determined to honour the Clark brothers by titling the new railway the "Ferrocarril Transandine Clark"—but forgot to do so. The first section from Santa Rosa de los Andes to Juncal opened in February 1906 and two years later the line was further extended to Portillo. In 1910 a further extension brought the line to a summit near the Argentine frontier where both sections were joined by the opening of the Cumbre (Summit) tunnel on 16 May that year when through traffic was inaugurated between Santa Rosa de los Andes and Mendoza. Boring of the 3463½yd Summit tunnel from Caracoles (in Chile), to Las Cuevas (in Argentina) was started from both ends in 1906. Originally Clark & Co. intended connecting the Chilean and Argentinian sections by a tremendous spiral tunnel of considerable length in order to avoid the worst summit snows. Fortunately a conventional tunnel was built beneath the Andes.

During construction of the Argentine Transandine Railway it was administered by the Argentine Great Western Railway until control of this entity passed to the Buenos Aires & Pacific Railway on 30 June 1907. Unfortunately the separate operation of the Chilean and Argentinian sections resulted in unsatisfactory service and the BA&PRly contract was cancelled. On 13 October 1921 the Argentine Government authorised the joining together of both the Argentine and Chilian Transandine Railways under a mutual agreement of the two Governments dated 3 December 1919. On 4 January 1922, a unification contract was signed, effective on 1 July 1923, by Samuel Hale Pearson (representing the Argentine Transandine Railway Co. Ltd.) of London, and J. Harry White representing the Chilian Transandine Railway Co. Ltd (also of London). The terms of the contract provided for the joint operation and administration of the railway, each company to retain receipts exclusive to its own system, and international traffic receipts to be apportioned on a stipulated basis. The Argentine Transandine Railway however reverted to BA&PRly control and in 1939 was purchased by the Government owned Argentine State Railways, now of course the Ferrocarril Nacional General Belgrano. The Chilian Transandine Railway Co. operated the Chilean section until it was handed over to the Chilian State Railways.

From an operation point of view, the Chilian Transandine Railway was divided into two sections: (1) Los Andes—Rio Blanco 21.14 miles and (2) Rio Blanco—Argentine Border 22.7 miles. The line was to prove extremely difficult to work containing severe gradients and sharp curves, not to mention additional hazards such as snow slides and avalanches. Heavy blizzards building up to 21ft of snow often brought traffic to a stop between June and October during the Andean winter. Locomotives were changed at Rio Blanco where most of the rack locomotives were stationed. CTR power ran through the Summit tunnel to Las Cuevas (3km further on) on the Argentine side where locomotives were again changed—those of the ATR taking over.

During the Andean winter from June—September, the mountains (and CTR) were liable to heavy snowfall—sufficient could fall in a day or a night to completely block cuttings and tunnel entrances. When the snow stopped a plough train would leave Rio Blanco consisting of an ordinary wedge plough pushed by two large rack/adhesion locomotives and a van carrying an extra gang of men. Experience proved that a wedge plough could usually get through to Juncal. The rotary plough was stationed at Juncal and replaced the wedge one at that station from whence it worked through to Caracoles (Chilean end of Summit tunnel), and on occasion passed through Summit tunnel to Las Cuevas. According to P. C. Dewhurst

(who worked on the line for a period) the worst section was between Km 55 and 62 where slides occurred filling up tunnel entrances with a mixture of snow and rocks. From about Km 62 to Caracoles and Las Cuevas the snow was fairly free from rocks, although it was quite common for there to be 14ft or more.

Because of the severity of the climb it was decided to utilise combined rack and adhesion working and part of the CTR was laid on the 3-bar rack system designed by Dr Roman Abt. This had been chosen on the recommendation of the Chilean Government Engineer Enrique Budge, and was eminently suitable for the small locomotives and light trains originally planned, but for 140 ton trains hauled by 90 ton locomotives the two bar system would have been more satisfactory. The terminus at Santa Rosa de los Andes lay 2669ft above sea level and in the 43.84 miles to the Argentine frontier, about two-fifths the way through the Summit tunnel, the line climbed 7783ft to 10,452ft although the actual summit of the Transandine Railway lay at 10,512ft within the tunnel. From Santa Rosa de los Andes to Rio Blanco (21.14 miles), a climb of

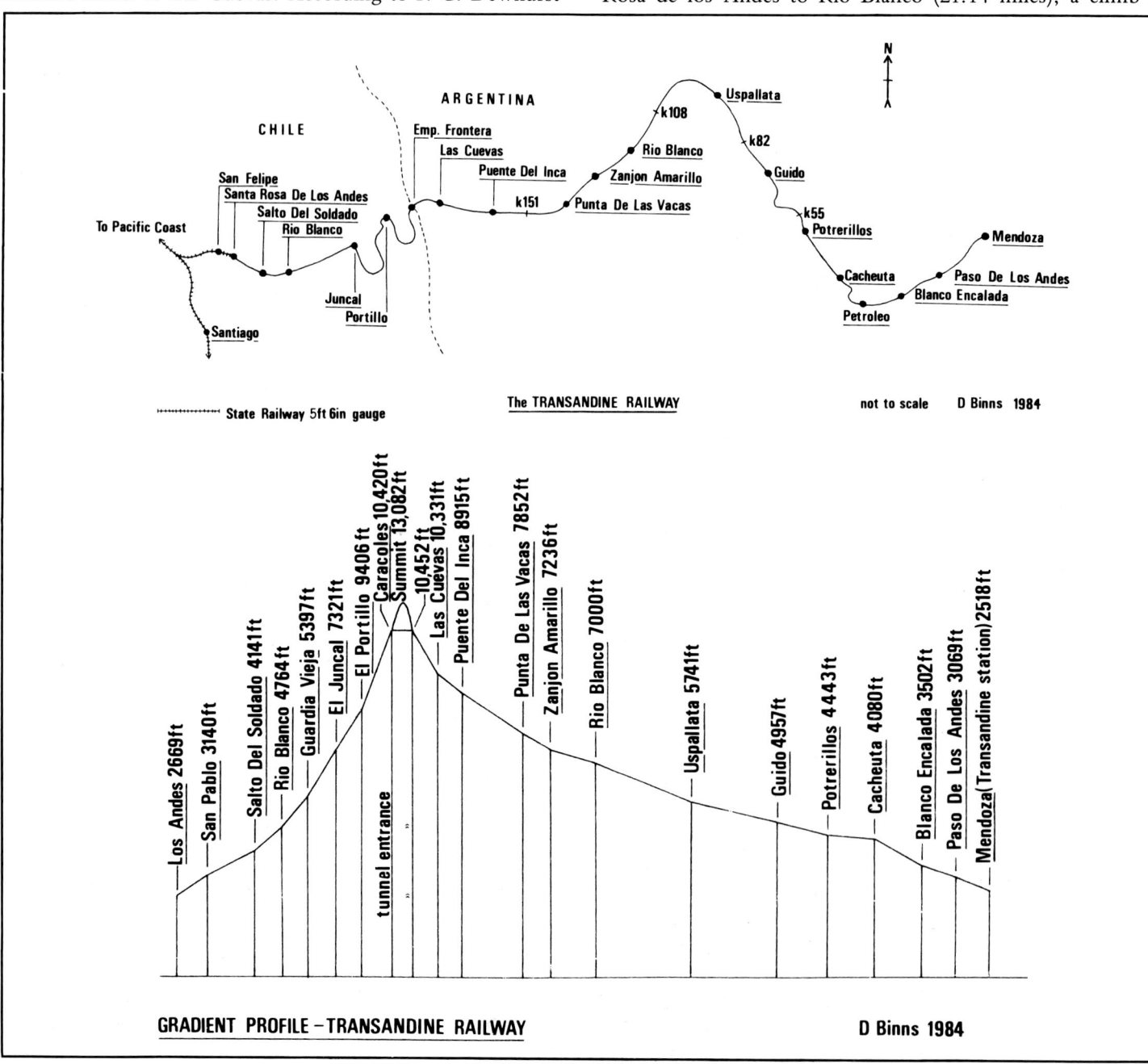

GRADIENT PROFILE-TRANSANDINE RAILWAY   D Binns 1984

2095ft was involved—this section being worked by Adhesion. The 22.7 miles from Rio Blanco to the Summit tunnel entrance on the Argentine side involved a total climb of 5688ft and rack and adhesion working was employed, Chilian locomotives running through the tunnel to Las Cuevas (some 3km further on) on the Argentine section where locomotives were changed. On the adhesion worked lower section of the CTR the ruling grade was 1 in 40 whilst the upper contained seven adhesion and six rack sections located as follows:

```
km 34.9−37.2 = 2.3km
   39.9−44.3 = 4.4
   46.3−49.2 = 2.9
   55.4−63.2 = 7.8
   63.4−64.9 = 1.5
   65.9−69.0 = 3.1
              ─────
              22.0km (13.67 miles)
```

Of this, 1¼ miles was at 1 in 12½, 2½ miles at 1 in 13.1 and 5 miles at 1 in 14.3 The Chilean section contained the longest continuous stretch of adhesion worked 1 in 40—22,782ft. Minimum radius on the adhesion section was 328ft except for one curve of 262ft, whilst on the rack it was 590ft. All these curves had the outer rail elevated by 3½in on adhesion sections and 2⅜in on rack sections. Both CTR and ATR adhesion sections were laid with 50lb rail on wood sleepers (Chilian Oak), but 55lb rail on steel sleepers (each weighing 119lb with fastenings) was employed for the rack sections due to the extra work thrown on the rail on steep grades by the action of the brakes. On the Chilean side were 21 large bridges (about 400 metres in total length) and 134 culverts and small bridges totalling some 180 metres. The two largest bridges were each 30 metres long. The CTR had 26 tunnels of which the longest was the Summit tunnel 10,391ft long.

On the CTR early train services were handled by 40 ton locomotives hauling 70 ton trains at 5mph on the rack, the Company purchasing two 2−6−2T and one 2−6−4T from A Borsig (of Berlin). Both classes had outside frames, rack engines and repression brakes. A Lima articulated 4+4+4 Shay was also tried and this could haul 80 tons up 1 in 12½ by adhesion alone but unfortunately with its 10mph top speed was too slow for use on the adhesion worked sections.

The metre gauge Argentine Transandine Railway extended from the city of Mendoza to the Chilean border, a distance of 111 miles. It climbed from 2518ft above sea level at the Mendoza Transandine station to 10,452ft at the Summit. The actual ascent of the Andes begins about 20 miles from Mendoza where the railway enters the gorge of the Mendoza river. From this point deep cuttings and numerous bridges and tunnels were needed and in 10 miles the railway crossed the river five times over bridges of spans varying from 148ft to 246ft. The ATR was not as steep as the Chilean section and contained 1 in 40 maximum adhesion grades to Punte de la Vacas. From this point to the summit were seven Abt rack sections on 1 in 16.6 grade totalling 8.7 miles. As with the CTR, adhesion sections were laid with 50lb rail on wood sleepers but 55lb rail on steel sleepers (each weighing with fastenings 119lb) was used for the rack section due to the extra work thrown on the rail on steep grades by the action of the brakes. By the mid-1920s 60 and 65lb/yd rail was in use. Minimum curve radius (adhesion) was 100 metres and on the rack 180 metres. An interesting point was that 34% of the entire length of the railway was made up of curves. There were 71 bridges and viaducts having a total length of 1442 metres, and ten tunnels having a total length of 2487 metres.

The ATR adhesion section from Mendoza to Punta de las Vacas was originally worked by six-coupled tank locomotives

Right: A Kitson Meyer articulated pushes the rotary plough through snow on the upper section of the Chilian Transandine Railway in an attempt to keep the line open. This photograph gives some idea as to the problems encountered during the average Andean Winter. C.J. Walker collection.

Below: A commercial card by Propiedad de J. Allan, Valparaiso. Train crew and passengers pose for their photograph to be taken on the Chilian section. Locomotive is rack/adhesion 2−6−4T No. 6 built by Borsig in 1906. Could this be the first express to leave Santa Rosa de los Andes for Juncal in 1906? Probably the two men and two boys with white aprons were the kitchen car and service staff. The locomotive is painted black and the coaches appear to be a reddish brown with white roofs. C.J. Walker collection.

Beyer Peacock rack/adhesion 0−6−2T (Lange & Livesey Patent) at work on the Argentine Transandine Railway. This type was designed by Beyer Peacock & Co. Ltd and the consulting engineers Livesey & Son in 1889 and a number of these locomotives were built in the period 1889-1899.

D. Binns collection.

built by Dübs & Co, whilst the upper rack section was handled by Beyer Peacock rack/adhesion four-coupled locomotives weighing 45 tons. With the Summit tunnel nearing completion a considerable increase in traffic was anticipated and the Consulting Engineers—Livesey Son & Henderson—were requested to investigate the motive power situation and to find a suitable type which could be used on both halves of the system. The Mallet type semi-articulated was rejected since it was considered impossible to design an adequate firebox and ash pan for 1 metre gauge. A Baldwin rack/adhesion design was rejected as unsuited for such rough work as was a rack/adhesion design from the Swiss Locomotive & Machine Works, Winterthur, who were of course experts in mountain railroading. The Beyer Garratt was considered but rejected as being unsuited to the rough work of clearing 20ft drifts with a wedge shaped plough. The requirements were bordering on the impossible—all other true mountain railroads only needing to haul one or two small coaches. This was not for the Transandine Railway who was planning through International express trains along with considerable freight. In the period 1910-1935 the Transandine was part of the only practical way of travelling by land from Buenos Aires to Valparaiso or Santiago. As part of these services the Company operated a Pullman car train comprising seven bogie coaches (built by the Birmingham Railway Carriage & Wagon Company). This consisted of four Pullman saloons, dining car, a combined kitchen/baggage car and a full baggage. It is recorded that the comfort of passengers had been a matter for particular study, the cars being "handsomely appointed and fitted with double windows and the most modern heating and lighting apparatus".

For these services the Transandine required locomotives capable of hauling 150 ton trains on the rack at 6.2mph uphill and 9.3mph down. Adhesion sections were scheduled at 18.6mph. The locomotives must be able to negotiate 328ft radius adhesion curves and 590ft rack curves and be able to work in the severe Andean winters which occurred between June and September and in which 20ft of snow could fall. In as little as 12 hours snow would completely fill in cuttings and block tunnel entrances which necessitated erection of numerous wood and galvanised iron snow-sheds. The requirements seemed imposible and in 1906 Livesey Son & Henderson approached Dr Roman Abt who conferred with the Maschienenfabrik Esslingen (of Esslingen, Wurtemburg). Eventually Abt reported that such a locomotive was impossible doubting if the track could withstand such stresses. Abt also maintained that the high tractive forces needed would cause bent frames on the rolling stock and so cause derailments. Both Dr Abt and Esslingen considered the provision of such a locomotive to be impossible advising that "terrible accidents much be expected if the Company tried to haul such loads". On receipt of Dr Abt's negative reply the Consulting Engineers were asked to design a suitable locomotive, eventually producing plans for an 0−8+6−0T combined rack/adhesion locomotive based on the Kitson Meyer form of articulation. The proposed locomotive had two steam bogies, the leading unit being an ordinary 4 axle adhesion bogie driven from its own pair of cylinders and capable of hauling the locomotive up the 1 in 12½ rack grades. The rear 3 axle steam bogie had two cylinders and inside frames which served as outside frames for the two sets of rack pinions located intermediately between each axle—this unit hauling the train. All 4 cylinders were outside and located at the outer ends of the units with balanced valves on top actuated by Walschaerts valve gear. At this stage the design was good but the CTR was concerned about loss of adhesion should the leading set of wheels slip on the 1 in 12½ grade. To overcome this a third rack pinion was mounted between the first and second axles and an extra pair of small cylinders was affixed on the adhesion bogie. The original design employed a larger boiler than was usually fitted to Kitson Meyer locomotives of that period but due to the additional auxilliary rack engine the boiler had to be reduced to keep within the 12 ton axle load. Livesey Son & Henderson did not approve of these additions but nevertheless the Company had the final word and the prototype emerged from the works of Kitson & Company in 1907 complete with extra cylinders and third rack pinion.

The Kitson Meyer rack/adhesion locomotives successfully hauled 150 ton trains but within a few months track was found to be displaced with many broken sleepers. The assistant Manager of the Chilian Transandine Railway, William Theodore Lucy, reported that in their original form this type of locomotive was unneccessarily complicated and later experience was to prove that the original plans produced by Livesey Son & Henderson, would have produced a far less complicated and more powerful and reliable machine. However not withstanding the warnings of Dr Abt, the Kitson Meyer locomotives successfully hauled the loads for which they were designed—a truly remarkable effort. The maximum load for these locomotives was 140 tons on the 1 in 12½, but in order to avoid straining either the rack or the pinion gearing on the locomotives, uphill passenger trains were limited to six bogie carriages weighing 120 tons. The Kitson Meyer's brought seven vehicles down the rack (140 tons) and in their original condition climbed the 1 in 12½ rack section on adhesion units alone, but the additional rack cylinders on the adhesion bogies proved too much for the boilers which could not make enough steam. In 1911 the small rack engine (which consisted of a single set of pinions driven by the pair of cylinders at the front

FCTC No. 7–the first Kitson Meyer rack/adhesion 0−8+6−0T (Kitson & Co. Works No. 4488/1907). The cylinders for the additional auxilliary rack engine can be seen above the cylinders of the leading unit. The short bunker/back tank are also evident in this official photograph. Note the cloth backdrop to this scene taken at the works.

D. Binns collection.

FCTC No. 8 (Kitson & Co. Works No. 4598/1908) was built new with bunker and back tank extended to the rear of the frame. It was photographed on the rack at km 48½ hauling a goods train on 1 August 1911 through mountain scenery typical of the Chilian Transandine.

P. C. Dewhurst via P. K. Dewhurst.

end of the adhesion bogie and mounted above the adhesion cylinders) was removed so returning these locomotives to the condition specified originally by the Consulting Engineers.

The Transandine Railway was one of the most unusual in the

A commercial postcard produced by Adolfo Conrads, Santiago showing a Kitson Meyer 0—8+6—0T on a goods train on the FCTC at Estacion Juncal 7321ft above sea level. The locomotive still has the extra pair of cylinders for the auxilliary rack engine so the date would be between 1907 and 1911. The house occupied by P. C. Dewhurst was the small stone building to the left of the large building with verandah (above the locomotive). The walls in his house were 20in thick and that was all the insulation they had plus a pot bellied stove for heating. P. K. Dewhurst collection.

whole world and because of the mountainous conditions was an extremely difficult road to work. When climbing the rack sections the Kitson Meyer locomotives worked bunker first at the front end of the train (unlike other mountain railways) allowing the driver to see any sudden rock or snow falls. Both the CTR and the ATR Kitson Meyer's had overhanging cab roofs in the form of a wide canopy over the windows and doors all the way round. On the original locomotives smoke in tunnels proved to be a problem but this was resolved by fitting backward deflecting cowls. The most difficult problem was not climbing the mountain section but lay in bringing down grade 120 ton trains, and the drivers needed to be expert in the use of the five different braking systems which were built into these locomotives. A Westinghouse automatic train brake was used for stopping the trains either normally or in an emergency by automatic action. This was not used to control the speed of descent on falling grades. A Westinghouse straight air brake was fitted on the locomotive and used high pressure air direct from the very large main air reservoir. This non-automatic brake was used to vary the speed of the train on rack sections. A Westinghouse non-automatic control brake was incorporated and was specially designed for controlling trains on the long falling grades and was used in combination with the ordinary automatic brake so that the latter was always available and held in reserve for emergencies such as breakage of couplings. The non-automatic control brake used straight air throughout the train and its power could be varied as required. This did away with the need to release the brakes on the train in order to re-charge the small reservoirs as was necessary with the ordinary automatic brake. The fourth system was the repression brake in which the locomotives cylinders are used as an air compressor. When using the repression brake the valve motion was put into gear opposite to the motion of the train and clean air was admitted through a special valve below the blast pipe outside the firebox, to the exhaust pipe which became the suction side. The steam inlet ports became the

outlet or feed to the steam chests and steam pipes which formed a reservoir for the compressed air as far as the regulator which was kept shut. Braking was controlled by a valve which allowed the compressed air to escape as necessary whilst water flowing by gravity was admitted into the cylinders in quantities sufficient to absorb the heat due to compression. Only enough water was used to produce a light cloud of steam from the chimney to which the outlet from the working valve was lead, together with a fine rain like spray. Too much water would cause a reduction in the pressure with consequential loss of brake power. The repression brake used the cylinders as an air compresor and the system permitted an easy graduation of brake power. The repression brake used on the adhesion and rack portions of the locomotive was independant, only the air inlet valve being in common. The reason for this was that the rack portion had to be released before leaving the rack whilst the adhesion portion was kept in use. The fifth brake was a band brake, but being unreliable, expensive to maintain and of little use, was removed about 1914. It comprised a 22in. drum which itself was unable to control the locomotive let alone the train as well. As a safeguard two air compressors were fitted to the Kitson Meyer locomotives to ensure a continuous supply of compressed air along with an extremely large Westinghouse air reservoir.

As we have noted the difficulties lay in bringing heavy trains down the rack and a descent rate of 4000ft per hour was found to be right having regard to passenger comfort. At a descent speed of 9.3mph it was possible to stop a train on the 1 in 12½ grade in only 105ft—this speed representing a descent of 4012ft per hour. Entering rack sections needed great care on the part of the driver to avoid damage to the entrance pieces or to the pinion. It was found that the best method was to run the adhesion unit at 3mph—just enough steam being given at the same time to the rack engine to cause the pinions to revolve slowly until they engaged with the rack. The pinions beneath the footplate were conveniently located to allow the driver to 'feel' the entry to the rack and in the event of mounting the rack, or if a derailment occurred, the train could be stopped quickly. Entry to each rack section needed skill, care and judgement and if the entry was not correctly effected could lead to breakage of the entrance pieces and rack bars. Each rack section was fitted with an entrance rail 10ft 9in long with specially shaped teeth, the rack bars being fixed on elliptical leaf springs allowing vertical flexibility to ease engagement of the pinions. The Kitson Meyer's were not turned at the top of the line and the firebox top and water gauges were designed for 1 in 12½ grades in the one direction only, the descent being made chimney first allowing the driver a better view of the line and having all the gauges and brake valves in front of him.

Really the Transandine Railway should have been a failure as everything was against it from gradients to weather, from complications of trying to work international expresses over a real mountain railroad by over complicated locomotives and the resultant operational difficulties. However the Kitson Meyer articulateds did the job well, the Chilian locomotives burning 'Crown Patent Fuel' briquettes in the early days, and later, coal. The Argentine locomotives used best Welsh coal. The average consumption of the Kitson Meyer rack/adhesion locomotives on the CTR in 1913, including lighting up, was 163lb per mile hauling average loads of 100 tons up and 140 tons down. This figure represented the total running—the uphill journey using 277lb of briquettes each mile. In addition to all the difficult operating problems the engine men had to contend with an extremely complicated set of controls amounting to 49 separate cab fittings:

    2 Regulator handles—adhesion and rack
    2 Reversing screws—adhesion and rack
    6 Pressure gauges: Boiler pressure gauge
        Adhesion engine steam pipe gauge
        Rack engine steam pipe gauge
        Westinghouse duplex gauge
        'Straight air' on engine gauge
        N A Control on train gauge
    1 Speed indicating gauge
    2 Water gauge glasses for boiler

1 Gauge for water in tanks
3 Air brake drivers valves: Automatic brake valve 'A'. Straight air valve for engine 'B'
N A Control valve for train 'C'
2 Repression brake valves—adhesion and rack 'D'
2 Hand brake screws—the hand brakes on the locomotive were connected with the same rigging which was operated by the straight air brake.

2 Band brake screws 'E'—later removed
1 Automatic brake isolating valve
2 Injectors—Gresham & Craven No. 9 Combination
2 Valves for filling repression water tank
1 Valve for water admission to cyinders
1 Valve (or lever) for repression air inlet
1 Detroit sight feed lubricator, 6 feeds
1 Steam valve to lubricator
1 Steam valve to adhesion cylinder cocks
1 Steam valve to rack cylinder cocks
2 Air pump steam valves RH and LH
1 Blower valve
2 Sanding levers or valves
1 Valve to boiler pressure gauge
1 Warming cock
1 Valve for automatic pinion lubricator (opens automatically the oil supply to the pinions whenever there is pressure of either steam or air in the steam pipe of the rack portion. Designed to lessen the drivers work and to avoid neglect, but also obviates all waste on the adhesion sections and ensures correct lubrication on the rack in both directions)
1 Whistle cord (or lever)
1 Headlight control—A small screw gear allowing the beam of the acetylene searchlight to be turned to right or left to suit curves
1 Acetylene gas generator
1 Smoke deflector lever—This operated a quadrant shaped hood which could be turned over the top of the chimney when required in tunnels to deflect smoke backwards

FCTC No. 9 was Kitson & Co. Works No. 4664/1909, built new with extended bunker and rear tank. In this photograph taken at Los Andes on 25 February 1910 the additional cylinders for the auxilliary rack engine can be seen. Notice the mixture of 4 wheel and bogie goods vehicles. No. 9 was to prove the lucky engine being today preserved in the outdoor museum in Santiago.
P. C. Dewhurst, via P. K. Dewhurst.

1 Oxygen Apparatus—For emergency use if the locomotive had to stop in a smoke filled tunnel. Comprised a steel flask with reducing valve and masks connected by flexible metallic tubing

1 Signalling bell—Used mainly for snow ploughing and when locomotives were coupled together

1 Rack Engine Speed Indicator—Allowing the driver to know the speed of the rack engine which was out of his sight—especially necessary to know the speed when entering the rack.

When new the Chilian 0−8+6−0T were also fitted with condensing apparatus and a direct acting feed pump on the right side of the tank to supply water to the boiler. This had been removed by 1914 when it was realised that where it was needed, i.e. in the tunnels on rack sections when the extra cylinders were exhausting, it was not possible to maintain steam without using all the available blast.

Kitson Meyer Rack/Adhesion Locomotives supplied to the Chilian Transandine Railway

0−8+6−0T Kitson & Co. Works No. 4488/1907. Steam trial 14/10/1907. Built new with short bunker and back tank. CTR No. 7. Later Chilean State Railways No. 3347. Scrapped in the 1960s.

0−8+6−0T Kitson & Co. Works No. 4598/1908. Steam trial 17/11/1908. Built new with bunker and back tank extended to the rear. CTR No. 8. Later Chilean State Railways No. 3348. Still working the Portillo−Las Cuevas section in 1963. In 1971 No. 3348 was repaired, and made a trial run to Las Cuevas with a regular freight train. The trial was not successful, due to the low speeds attained, poor steaming (probably due to low quality coal being used) and problems with smoke in the tunnels. No. 3348 remained at Los Andes and remained in tact except for her front number plate and whistle which have probably been taken by somebody who once worked with her. On 6 September 1979 she was seen mounted on a plinth at Los Andes as FCTC No. 3348. The Chilean Railway Conservation Association (ACCPF) inspected the locomotive in 1985 and found that she could be repaired to an operational status given approximately 15 new boiler tubes and renewed firebox stays. The work was not

carried out and at the time of writing (February 1993) the locomotive remains in the roundhouse at Los Andes. Latest report from Ian Thomson states that No. 3348 is kept in very good condition and the ACCPF are again thinking about repairing it for use on excursion trains.

0−8+6−0T Kitson & Co. Works No. 4664/1909. Steam trial 5/7/1909. Built new with bunker and back tank extended to the rear. CTR No. 9. Later Chilean State Railways No. 3349. Still working the Portillo−Cuevas section in 1963. At Los Andes 1972—retained for snow clearance. Now preserved at the Parque Ferroviario, Quinta Normal, Santiago, and is on display in this outdoor garden museum.

All three Chilean Kitson Meyer's were in service on 30 June 1928 for passenger and goods workings despite a decree issued on 5 August 1925 authorising the CTR to electrify its line according to a plan submitted by Charles Wilson, and the technical advisor of the line, J. J. Fifer, which was approved by both the Company and the Chilian Government. On 29 October 1927 the first part of the electrification scheme opened to traffic but completion of the electrification from Rio Blanco to Las Cuevas in Argentina, was not completed until 1942. Power was supplied by the Compania Chilean de Electricidad, of Santiago. This was three phase, 50 cycle power at 44,000 volts supplied to an automatic sub-station at Juncal and two 1,500 motor generators supplied the catenary. Energy was received by the railway at Los Andes and the most convenient system of traction was a continuous current of 3,000 volts which was the same as that used on the State Railways. Because of this, through working was possible between the terminals of the State Railways and the CTR. Initially three electric locomotives were purchased from Switzerland.

Dr. Roman Abt noting that the Kitson Meyer locomotives were doing the job which he had pronounced impossible, subsequently approached the CTR offering to construct suitable rack/adhesion locomotives and the Company purchased two 0−6−8−0T semi-articulated units from the Maschienenfabrik Esslingen. In service these were not satisfactory mainly because they lacked the flexibility of the Kitson Meyer's which could pass from one curve to another with only 12ft or so of straight track between. The Kitson Meyer units were better on snow clearance and it was found easier to re-rail a Kitson than

Chilean State Railways 0−8+6−0T (Kitson Works No. 4664/1909) No. 3349 at Los Andes in 1977 in final guise.   A. E. Durrant.

Right hand rear view of FCTC No. 3349 at Los Andes in 1977.
A. E. Durrant.

an Esslingen. A definite advantage of the Kitson Meyer was that the steam bogies could be disconnected and run clear of the superstructure in three hours, the Esslingen requiring two days for this operation. This work was carried out on the Kitson Meyer's in the running sheds but the Esslingen had to be taken into the works before its bogie could be disconnected. All in all the Kitson Meyer was a far better design having regard to maintenance and access to boiler and firebox.

0-6-8-0T. One built by Maschienenfabrik Esslingen 1908. Works No. 3477. FCTC No. 10 (at first numbered 9 but later renumbered 10). Later Chilian State Railways No. 3350 at some unknown date after 1934. Worked until 1963 on the Portillo—Las Cuevas section but its rack gear was incorrectly handled by its driver and seriously damaged. It was hauled dead to Los Andes where it was stored for a few years before being cut up on the spot. Believed to be at Rio Blanco 1977.

A second Esslingen semi-articulated followed and this incorporated alterations to the steam pipes, side tanks partially extended over the front cylinders, and had two Westinghouse air compressors.

0-6-8-0T. One built by Maschienenfabrik Esslingen 1911 Works No. 3623. FCTC No. 11. Originally fitted with condensing apparatus and a direct acting feed pump—but the condensing apparatus was later removed. Later Chilian State Railways No. 3351 at some date after 1934. Possibly written off in 1940s accident.

Principal dimensions of this pair were:

Cylinders—adhesion	15⅜in×19¾in (rear unit)
Cylinders—rack	21¼in×17¾in (front unit)
DW diameter—leading	2ft 5½in
DW diameter—rear	2ft 11¾in
BP	210lb
HS tubes	2066sq ft
HS firebox	119sq ft
Grate area	34½sq ft
Water	1980 gall
Coal	2½ tons

Wt in W.O. No. 10— 81 tons. No. 11-85 tons
Westinghouse and hand brakes, all wheels. Also special hand operated band brakes on rack pinion drums.

Esslingen 0—6—8—0T No. 11 in service on the Chilian Transandine Railway.
D. Binns collection.

Four different views of Rack/Adhesion Kitson Meyer 0–8+6–0T FCTC No. 3349 as it looked in 1987 preserved in the Parque Ferroviario, Quinta Normal, Santiago.

Three more views of FCTC No. 3349 at Quinta Normal in 1987.

The four photographs opposite and the three larger ones on this page were taken by a gentleman in the USA. Unfortunately we cannot trace his name and address and regret therefor the lack of photographic credit. If the owner of these photographs reads this, would he please make contact.

Cabside plates of FCTC No. 3349 at Los Andes in 1977.
Top to bottom. "3349", "Kitson & Co. Limited, 1909, Leeds", "Reformada en Los Talleres del F.C.T.C., 1914, Los Andes".
                                    A. E. Durrant.

The other side of No. 3348 at Los Andes on 6 September 1979.

K. R. Chester.

FCTC Kitson Meyer No. 3348 on display in the station at Los Andes 12 October 1987. An exhibition of Transandine equipment was shown in the station for the benefit of the passengers of a special train from Santiago, organised by the ACCPF.

I. Thomson.

TRANSANDINE RAILWAY KITSON MEYER LOCOMOTIVE HAULING CAPACITY AT 8–10 mph

Wheels	0–8 +6–0T	0–8 +6–0T
Maker	Kitson	Kitson
Works No.	4674	4882/3
On level track – tons	2602	2595
On 1 in 100 – tons	619	612
On 1 in 75 – tons	418	473
On 1 in 50 – tons	318	311
On 1 in 25 – tons	131	124
On 1 in 12½ rack – tons	182	120

Kitson Meyer Rack/Adhesion Locomotives supplied to the Argentine Transandine Railway

0−8+6−0T Kitson & Co. Works No. 4669/1909. Steam trial 16/7/1909. FCTA Class E22. Road No. 38−later No. 41. Seen working in Mendoza yard of the Central Belgrano Railway in 1972.

0−8+6−0T Kitson & Co. Works No. 4670/1909. Steam trial 27/7/1909. FCTA Class E22. Road No. 39−later No. 42. Seen dumped at the Works of Tafi Viejo (nr Tucuman) 1972.

0−8+6−0T Kitson & Co. Works No. 4674/1909. Steam trial 12/10/1909. FCTA Class E22. Road No. 40−later No. 43? See note below. Seen dumped at Tafi Viejo 1972.

0−8+6−0T Kitson & Co. Works No. 4842/1911. Steam trial 16/5/1911. FCTA Class E22. Road No. 43? See note below. Seen dumped at Tafi Viejo 1972.

Note 1: There is some doubt as to the final road number of Works No's 4674 and 4842. No. 4674 may not have been renumbered 43 and could have retained 40 throughout its existence, in which case No. 4842 carried No. 43. This appears to be confirmed by a works list brought back from a 1972 visit.

0−8+6−0T Kitson & Co. Works No. 4882/1912. Steam trial 3/9/1912. FCTA Class E23. Road No. 44. Seen withdrawn and dumped at Tafi Viejo 1972.

0−8+6−0T Kitson & Co. Works No. 4883/1912. Steam trial 3/10/1912. FCTA Class E23. Road No. 45. Seen reputedly withdrawn at Tafi Viejo 1971. On the occasion of a visit in 1972 this locomotive was not seen but was supposed to be working at Mendoza.

One Kitson Meyer was seen derelict at Cordoba, Argentina in 1970. In 1988 Kitson Meyer No. 40 in pretty good condition, was locked away in a run-down shed at Tafi Viejo and carried no builders or number plates. No photographs were allowed, the manager being afraid to in case his superior would find out. (This information was provided by John Verser (USA) 30/11/1989).

In its 154½ miles the Transandine Railway had nine tunnels on the Argentine section with a total length of 575 yds, and 26 on the Chilian section which amounted to 3481 yds. Including the summit bore the total yardage for tunnels was 7519½ and it is no small wonder that one problem encountered by the operating departments was that of maintaining a breathable atmosphere in the locomotive cabs whilst working hard climbing the steeply graded track which pierced many tunnels. On several occasions the air became foul due to the poor quality Chilian coal then in use, resulting in the collapse of the crew. The Kitson Meyer locomotives were of course fitted with oxygen apparatus for use in such cases.

Because of the poor quality local coal and the high cost of imported Welsh Coal, the CTR electrified the section from Rio Blanco to Las Cuevas in 1927 but the ATR did not electrify.

On the Argentina side the ATR was cut for ten years (1934-1944), the cause being the collapse of an ice wall which released the contents of a large lake into the Mendoza river, swelling it to unbelievable size and ferocity. The ATR track was destroyed for miles on end and rail services from Chile could only operate as far as Punta de Vacas, where connection with Mendoza was made by road. In 1944 through rail service were re-introduced over a reconstructed road-bed.

As late as April 1989 the Transandine Railway was still working from Mendoza to Potrerillos (a distance of about 70km). There were four trips every weekend and for this job Ganz diesel railbuses were provided. Transandine may eventually reach to Uspallata, but from Uspallata to Las Cuevas has been cut in some places by "nature". From Las Cuevas to Los Andes was in very bad condition, many sections were cut by avalanches, or man removed. The Catenary was almost gone but in 1976 a very nice Swiss electric loco with rod transmission was seen in blue livery but this has not been seen since. By 1992 much of the Catenary was down and the poles broken off by the snow, some snow sheds have also been damaged and it is doubtful if the Argentina section will ever be repaired.

TRANSANDINE RAILWAY KITSON MEYER LOCOMOTIVE DIMENSIONS

Railway	CTR	CTR	CTR	ATR	ATR	ATR	ATR
Gauge	1m.	1m.	1m.	1m.	1m.	1m.	1m.
Wheels	0–8+6–0T	0–8+6–0T	0–8+6–0T	0–8+6–0T	0–8+6–0T	0–8+6–0T	0–8+6–0T
Maker	Kitson	Kitson	Kitson	Kitson	Kitson	Kitson	Kitson
Works No.	4488	4598	4664	4669/70	4674	4842	4882/3
Year	1907	1908	1909	1909	1909	1911	1912
Cylinder position	Outer	Outer	Outer	Outer	Outer	Outer	Outer
Cylinders – large rack – inches	18 × 19	18 × 19	18 × 19	18 × 19	18 × 19	18 × 19	18½ × 19
Cylinders – small rack – inches	13 × 14	13 × 14	13 × 14	13 × 14	13 × 14	13 × 14	–
Cylinders – adhesion – inches	16½ × 19	16½ × 19	16½ × 19	16½ × 19	16½ × 19	16½ × 19	16½ × 19
Boiler Pressure – lb/sq.in.	200	200	200	200	200	200	200
HS Firebox – sq.ft.	130.24	130.24	130.24	130.24	130.24	130.24	140
HS Tubes – sq.ft.	1767.76	1767.76	1767.76	1767.76	1767.76	1767.76	1910
HS Total – sq.ft.	1898	1898	1898	1898	1898	1898	2050
Grate area – sq.ft.	31	31	31	31	31	31	34
Driving wheel diameter	3ft.	3ft.	3ft.	3ft.	3ft.	3ft.	3ft.
Rigid wheelbase							
1st group	10ft. 6in.	10ft. 6in.	10ft. 6in.		10ft. 6in.	10ft. 6in.	10ft. 6in
2nd group		8ft. 11in.	8ft. 11in.			8ft. 11in.	8ft. 11in.
Total wheelbase	31ft. 2½in.	31ft. 2½in.	31ft. 2½in.	31ft. 2½in.	31ft. 2½in.	31ft. 2½in.	31ft. 10½in.
Water – gallons	1800	1800	1800	2100	2100	2212	2212
Coal – tons	2½	2½	2½	2½	2½	2½	3½
Weight empty – tons/cwt.			82–8				
Weight in working order – tons/cwt.	89–6	89–6	89–6	90–16	90–16	92–16	96–14
Weight adhesive – tons/cwt.	50–13	50–13	50–13	50–13	50–13		54–0
Tractive effort							
Adhesion engine lbs.					21,525 (75%)	21,530 (75%)	21,525 (75%)
Large Rack lbs.					34,186 (75%)	34,186 (75%)	36,150 (75%)
Small Rack lbs.					16,883 (75%)	16,883 (75%)	None
Height – overall							
Width – overall			9ft. 5¼in.				
Length – over buffer	46ft 11in	46ft 11in	46ft 11in				

FCTA 0–8+6–0T Kitson Meyer Rack/Adhesion locomotive No. 40 as built. Kitson Works No. 4674/1909.   *D. Binns collection.*

The Argentine Transandine Railway became part of the Ferrocarriles Nacionales General Belgrano (EFEA). No. 41 was photographed at work in Mendoza yard in 1972 and over the years had been substantially altered including removal of the auxilliary rack engine and cylinders, and removal of its condensing apparatus. Longer side tanks have been fitted, a second dome added and a generator and electric lighting. The canopy roof has been replaced by one of a more conventional outline and

the bunker has been built up.
D. Ibbotson.

Argentine Transandine Railway 0−8+6−0T No. 42 as built—photographed in the snow on a passenger train in 1910. P. C. Dewhurst via P. K. Dewhurst.

No. 42 was seen on the General Belgrano Railway on 21 April 1971. It too had been altered along the lines described in the top caption on this page. For its appearance as built see the photograph above.
D. Trevor Rowe.

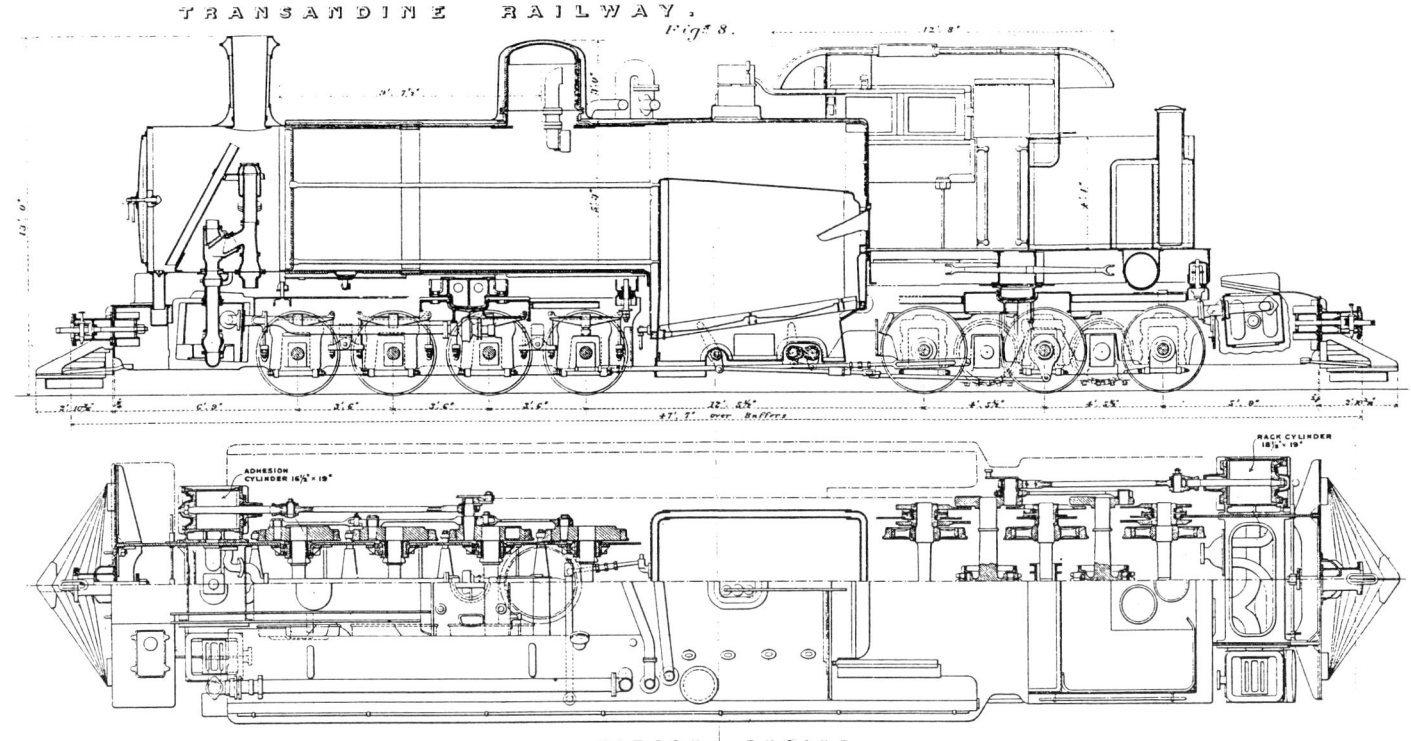

TRANSANDINE RAILWAY.
Fig.ª 8.

KITSON ENGINE.

44

The final pair of 0−8+6−0T emerged from the works of Kitson & Co. in 1912, Works No's 4882/3 for the Argentine Transandine No's 44 and 45. These were built new without the additional rack engine on the leading adhesion unit and were thus provided with full length side tanks. This pair had inside frames for the front adhesion bogie and outside for the rear rack unit.

D. Binns collection.

FERRO CARRIL DE BUENOS AYRES AL PACÍFICO.

THROUGH TIME TABLE

From 15th December, 1906, till further notice.

	Monday, Wednesday, Friday.		
Buenos Ayres (Retiro) dep.	7.00 a.m.		
Mercedes	9.05 a.m.		
Junin	11.35 a.m.		
Rufino	2.25 p.m.	Buenos Ayres and Pacific and Argentine Great Western Railways Through Train. Dining and Sleeping Cars.	Lunch, Dinner and early Coffee on Train, $5.00 to $6.00
Laboulaye	3.30 p.m.		
Mackenna	5.10 p.m.		
Villa Mercedes arr.	8.20 p.m.		
Villa Mercedes dep.	8.30 p.m.		
(Sleep in Train)			
San Luis	11.05 p.m.		
	Tuesday, Thursday, Saturday.		
Mendoza (Change) arr.	4.30 a.m.		
Mendoza dep.	5.00 a.m.		
Puente del Inca arr.	11.00 a.m.		
Las Cuevas	12.00 noon Chile Time.		Lunch, $2.00 to $3.00
(Change Rail to Coach and Lunch)		Transandine Railway (Buffet on Train.)	
Las Cuevas dep.	12.00 noon		
Summit arr.	1.20 p.m.		
Juncal	4.00 p.m.		
(Change Coach to Rail)			
Los Andes arr.	6.15 p.m.		Dinner - $2.00 (at Los Andes or Llai-Llai.)
(Customs and Change)			
Los Andes dep.	6.50 p.m.	Chilian Transandine Railway.	
Llai-Llai arr.	8.30 p.m.		
Llai-Llai dep.	8.50 p.m.		
(Change)			
Valparaiso arr.	10.40 p.m.	Chilian State Railways.	
Santiago	10.15 p.m.		

APPROXIMATE COST OF JOURNEY—BUENOS AYRES TO VALPARAISO OR SANTIAGO.

Through Fares:—including Beds in Train, 50 kilogrammes Luggage, collected from and delivered to Hotels at points of departure and destination.

First Class	£13. 10s. 0d.	
Second "	£8. 10s. 0d.	
Half Ticket (3 to 12 years)	£8. 0s. 0d.	
	£5. 0s. 0d.	

ENQUIRY OFFICES:
EXPRESO VILLALONGA, SANTIAGO, Moneda, 944.
VALPARAISO, Blanco 241 and Cochrane 26.

RETURN JOURNEY.
CHILE TO BUENOS AYRES.

From 15th December, 1906, till further notice.

		Monday, Wednesday, Friday.	
Santiago	dep.	6.15 p.m.	Chilian State Railways.
Valparaiso	"	5.00 p.m.	
Los Andes	arr.	10.00 p.m.	
(Change)			
		Tuesday, Thursday, Saturday.	
Los Andes	dep.	5.00 a.m.	Chilian Transandine Railway.
Juncal	arr.	8.00 a.m.	
(Change)			
Juncal	dep.	8.30 a.m.	Coach.
Las Cuevas	arr.	2.00 p.m.	
(Customs and Change)		(Argentine Time.)	
Las Cuevas	dep.	2.30 p.m.	
Puente del Inca	arr.	3.20 p.m.	Transandine Railway.
Mendoza	"	9.00 p.m.	
(Change)			
Mendoza	dep.	10.00 p.m.	
(Sleep in Train.)			
		Wednesday, Friday, Sunday.	Buenos Ayres and Pacific and Argentine Great Western Railways Through Train. Restaurant and Sleeping Cars.
Villa Mercedes	dep.	6.50 p.m.	
Buenos Ayres	arr.	6.50 p.m.	

For those desirous of spending one day in Mendoza, the through trains leaving Buenos Ayres on Tuesdays, Thursdays and Saturdays are convenient. These trains start at the same time as above, but there is no connection on the Transandine Line until that shown above on the following day.

NOTE.—The $ are paper dollars, and subject to fluctuations in value; at present Argentine dollars are $11.45 per £.

The cost of this journey is the same as that shown on page 9. In addition to the above trains there are Through Expresses leaving Mendoza at 6 p.m. for Buenos Ayres on Mondays, Wednesdays and Fridays.

KITSON MEYER LOCOMOTIVES IN OTHER PARTS OF THE WORLD

SOUTH AFRICA

Four Kitson Meyer locomotives were tried in South Africa, two on Rhodesia Railways and one each on the Central South Africa and Cape Government lines. All were 0−6+6−0 of similar appearance with separate bogie tenders attached to the rear of the locomotives. No side tanks were provided and all had their cylinders at the rear of each steam bogie. All four units were similar in size but there were dimensional differences. Wiener in his book *Articulated Locomotives* records an additional locomotive—a solitary Kitson Meyer 0−6+6−0T as being supplied by Kitson & Co. in 1904 to the Central South African Railway and he also quotes dimensions. Extensive research has not turned up any evidence of this locomotive and it cannot be traced from the surviving records of Kitson & Co. Experts in South African locomotive history have also drawn a blank pointing out that the small coal and water capacity of a Kitson Meyer tank locomotive would be totally unsuited to South African conditions. It seems certain that Wiener is in error and this locomotive did not exist.

The first Kitson Meyer was supplied to South Africa in 1903 to the 3ft 6in gauge Cape Government Railway. This unit worked out of East London on the Cape Eastern section until 1908, being scrapped by South African Railways in either 1910 or 1912.

Kitson Meyer locomotive supplied to the Cape Government Railway

0−6+6−0. Built by Kitson & Co. 1903. One locomotive supplied. Works No. 4197. Type 2 simple locomotive with the cylinders at the rear of each unit and with a separate tender. CGR Road No. 800. Steam trials 26/8/1903.

The second and third Kitson Meyer tender locomotives were supplied to the Rhodesia Railways in 1903.

The natural port for the Rhodesias was Beira in Portugese East Africa and it was decided to connect Beira with Salisbury, the administrative capital of Southern Rhodesia. The new line

was constructed by the Beira Railway Company and this included a steep climb from the low lying coastal area up to 4,000ft level of the bulk of the Rhodesias. The hill section up to Umtali, principal town of the (then) eastern districts of Southern Rhodesia, opened in 1896, the line starting some 36 miles inland from Beira at a place called Ponte de Pungwe. In 1898 an extension opened from Ponte de Pungwe over difficult swampy ground to Beira—the whole route from Umtali to Beira being laid to 2ft 0in gauge. Quite soon it was found that this narrow gauge restricted traffic capacity and in 1900 the line was converted throughout to 3ft 6in gauge. A separate Company—the Mashonaland Railway—was formed to continue the line from Umtali to Salisbury—capital of the British South Africa Company Territory of Rhodesia, this being laid to 3ft 6in gauge from the start, and opening in 1899. The Beira Railway remained a separate entity due to its geographical location but was worked as an integral part of the Rhodesia Railways. In later years the Mashonaland Railway and the Rhodesia Railways amalgamated but previous to this had been worked by one central management.

The main line was single track and laid with 60lb rail which allowed a 13½ ton maximum axle loading. From Beira to Vila Machado, the route was comparatively easy but for 143 miles between Vila Machado and Umtali, the line climbed from 185ft to 3552ft above sea level. The ruling grade was 1 in 50 uncompensated for 330ft unchecked curves which is equal to approximately 1 in 37, and it was only after some years that check rails were installed. There was considerable mineral traffic to the port of Beira and during the early years of this century Neilson Reid 4−8−0 locomotives worked the Beira−Umtali section. Due to the severity of the grades heavy freight trains had to be run in two sections and as traffic increased larger motive power became necessary resulting in an order being placed for two Kitson Meyer articulated locomotives. These were despatched from the Leeds works of Kitson & Co.

Cape Government Railway Type 2 Kitson Meyer 0−6+6−0 with separate tender was built by Kitson & Co., Works No. 4197.   D. Binns collection.

The Cape Government Railway 0–6+6–0 became No. 800 and was photographed possibly at de Doorns where it may have been used for banking.
Photographer unknown, via A. E. Durrant.

The opposite side of CGR No. 800.
Uwe Bergmann collection.

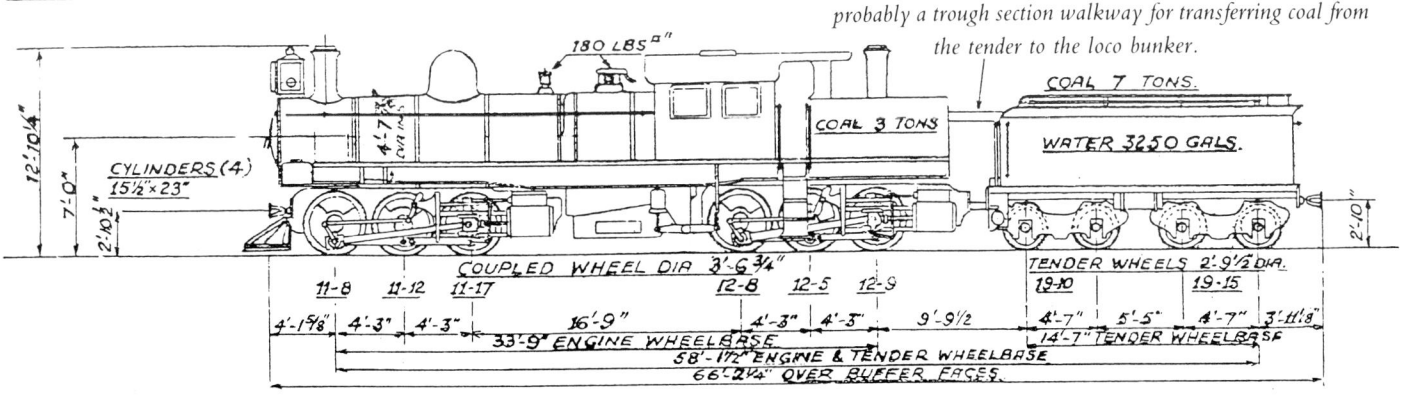

HEATING SURFACE TUBES 1590 SQ. FEET
" " FIREBOX 133 " "
" " TOTAL 1723 " "
FIREGRATE AREA 33 " "
234 TUBES 1⅞" EXT. DIA.
13'-9 ⅝" BETWEEN TUBEPLATES

TRACTIVE FORCE 75% 34,900 LBS
MAKERS KITSON & CO - 4240/1
ENGINE Nº 51 & 52
DATE IN SERVICE 1904
SCRAPPED 1912

CLASS KM
KITSON-MEYER ARTICULATED
RHODESIA RAILWAYS

probably a trough section walkway for transferring coal from the tender to the loco bunker.

COAL 7 TONS.

WATER 3250 GALS.

COAL 3 TONS

180 LBS

12'-10¼"

7'-0"

2'-10¼"

CYLINDERS (4)
15½" x 23"

4'-7"

2'-10"

COUPLED WHEEL DIA 3'-6¾"

TENDER WHEELS 2'-9½" DIA.

11-8   11-12   11-17   12-8   12-5   12-9   19-10   19-15

4'-1⅝" | 4'-3" | 4'-3" | 16'-9" | 4'-3" | 4'-3" | 9'-9½ | 4'-7" | 5'-5" | 4'-7" | 3'-11⅛"
33'-9" ENGINE WHEELBASE   14'-7" TENDER WHEELBASE
58'-1½" ENGINE & TENDER WHEELBASE
66'-2¼" OVER BUFFER FACES

TOTAL WEIGHTS   ENGINE 71 TONS 19 CWT   TENDER 39 TONS 5 CWT   WORKING ORDER.

in parts, being assembled at the Umtali works and entering service in 1904. They were not a success and spent a lot of time undergoing heavy repairs caused by excessive flange wear, probably brought about due to the drive being taken to the leading wheels of each steam bogie. In 1907 both Kitson Meyer locomotives were transferred to coal haulage on the difficult section between Wankie Colliery and Bulawayo, which had opened in 1903. This was laid with 80lb rail on wood sleepers and from Dett (3,590ft) the route descended to Wankie (2,448ft) in 51 miles at 1 in 61.3 ruling, compensated for 525ft curves. The Wankie Colliery Company supplied large quantities of coal to the railways, power and electricity supply stations, and to the copper mines. On this work the Kitson Meyer's proved heavy on coal and water and were disliked due to their excessively hot cabs and low operating speed of 8–10mph.

Kitson Meyer locomotives supplied to Rhodesia Railways 0–6+6–0. Built by Kitson & Co. 1903. Two locomotives supplied. Works No's 4240/1. Type 2 simple locomotives with the cylinders mounted at the rear of each unit and with separate tenders. Road No's (Rhodesia Railway) 51 and 52. Steam trials 23/11 and 10/12/1903. Withdrawn 1910 and dismantled two years later. The boilers were still in use at the Umtali shops in 1928.

One Kitson Meyer locomotive similar to the Rhodesian units was delivered by Kitson & Co. in 1904 to the 3ft 6in gauge Central South African Railway. This had its cylinders at the rear of each steam bogie and a separate tender was trailed at the rear of the locomotive. No side tanks were provided. The CSAR unit entered service in 1904 at Germiston and was used on Reef services, most probably on coal haulage. Before the Union there were three main railways in South Africa—the Cape Government and the Natal Government working in their respective provinces and the Central South African which covered the Orange Free State and the Transvaal. As far as can be traced, this Kitson Meyer worked in the Transvaal until shortly after 1910 when South African Railways were formed by amalgamation. It was sold in 1918 to the Transvaal & Delagoa Bay Collieries in the East Rand area but its subsequent history is not known.

Railway	Rhodesia	CSAR	Cape Govt.
Gauge	3ft 6in	3ft 6in	3ft 6in
Wheels	0–6+6–0	0–6+6–0	0–6+6–0
Maker	Kitson	Kitson	Kitson
Works No.	4240/1	4262	4197
Year	1903	1904	1903
Cylinder Position	rear	rear	rear
Cylinders—inches	15½ × 23	16 × 24	16 × 24
Boiler Pressr.—lbs/sq in	180	180	180
Heating Surface:			
Firebox—sq ft	133	136	133
Tubes—sq ft	1590	1727	1590
Total—sq ft	1723	1863	1723
Superheater—sq ft	None	None	None
Total HS—sq ft	1723	1863	1723
Grate area—sq ft	33	34	34
Driving wheel—dia	3ft 6¾in	4ft 0in	4ft 0in
Other wheels—dia	None	None	None
Rigid wheelbase—			
first group	8ft 6in	8ft 6in	8ft 6in
second group	8ft 6in	8ft 6in	8ft 6in
Locomotive:			
Total wheelbase	33ft 9in	34ft 0in	34ft 0in
Water—galls	None	None	None
Coal—tons	3	3	3
Wt. empty—tons/cwt	71 19	83 3	81 18
Wt. in W.O.—tons/cwt	71 19	83 3	81 18
Wt. Adhesive—tons/cwt	71 19		
Tractive effort—lbs	34,900 (75%)		
Height—overall	12ft 10¼in	12ft 10in	
Width — ”			
Length— ”	66ft 2¼in	66ft 5¼in	
Tender:			
Wheel—dia	2ft 9½in	2ft 9½in	
Wheelbase total	14ft 7in	14ft 7¼in	
Water—galls.	3250	3000	3250
Coal—tons	7	7	7
Wt. in W.O.—tons/cwt	39 5	38	
Wheelbase—loco/tender	58ft 1½in	58ft 4½in	

Rhodesia Railways Kitson Meyer 0–6+6–0 Type 2 with separate tender. No.52 was photographed at Wankie circa 1907.   A. H. Croxton.

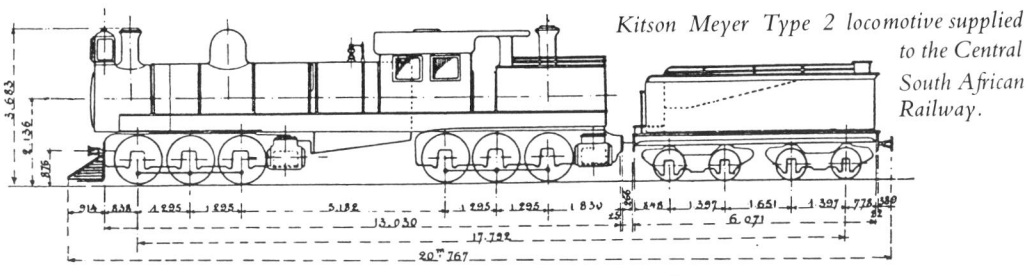

Kitson Meyer Type 2 locomotive supplied to the Central South African Railway.

Kitson Meyer locomotive supplied to the Central South African Railway

0−6+6−0. Built by Kitson & Co. 1904. One locomotive supplied. Works No. 4262. Type 2 simple locomotive with the cylinders at the rear of each unit and with a separate tender. CSAR Road No. 1000. Steam trials 7−9/4/1904. Became SAR No. 1600.

Map courtesy "Railway Magazine" November 1940.

J. Agnew collection.

Thought to be CSAR No. 1000 and later No. 1600, this Type 2 Kitson Meyer 0−6+6−0 has its tender at the opposite end to usual. It is thought to be Works No. 4262/1904 but its number plate only appears to have two digits (No. 30 or 36?)−could this have been its Transvaal & Delagoa Bay Collieries number?

Central South Africa Railway 0−6+6−0T No. 1000−later renumbered No. 1600−was photographed in Johannesburg station prior to April 1918 when it was sold to the Transvaal & Delagoa Bay Collieries, Witbank.

South African Railways.

JAMAICA

The Jamaica Railway was noteworthy as being the earliest colonial railway, the first section between Angels (2½ miles beyond Spanish Town) and Kingston, a distance of some 14½ miles, opening on 21 November 1845. One hundred years on, the tracks of the Jamaica Government Railway had expanded to 216 miles. In 1867 the line had been extended to Old Harbour and in 1879 the Government bought the railway and further extensions were then put in hand. Expansion proved difficult due to the mountain ranges which ran from east to west and also a series of lesser mountains some running parallel to the main range, others running in south-easterly and north-westerly directions. The highest peak was some 7,400ft above sea level in the Blue Mountains and these ranges were to prove an effective barrier to the railway contractors. On 1 January 1890 the Jamaican Government sold the railway to the West Indian Improvement Company in return for £800,000 and an undertaking that it would extend the lines to Montego Bay and to Port Antonio at its own expense. The West Indian Improvement Co. honoured its obligations but went bust in the process —the railway reverted to Government ownership in April 1900.

By 1918 the 4ft 8½in gauge Jamaican Government Railway extended some 197½ miles with two main lines, the first from Kingston to Port Antonio and the second Kingston to Montego Bay.

The railway played an important part in Anglo-American defence with US air and naval bases rail connected. Gradients on much of the line were severe with 1 in 30 ruling, the summit on the Montego Bay line being at an elevation of 1,700ft above sea level. The summit on the Port Antonio line was at 750ft. The route was laid with 60/80lb rail on hardwood sleepers and the ruling curvature on the mountain section was 300/330ft. These heavy grades and sharp curves limited the loadings on the mountain sections to 5 loaded 40 ton double bogie wagons, whereas on the more level sections locomotives could haul an average of 30 loaded 40 ton double bogie wagons.

Over the hill sections passenger trains averaged 18mph and goods 10—15mph. The principal freight traffic was alumina and alumina processing materials, sugar cane and bananas. Locomotives and stock were mostly of American design although some of the motive power was British built. During 1904 Kitson & Company supplied three Kitson Meyer articulated locomotives, these entering service in the following year. The girder main frame carrying the taper boiler, cab, water and fuel compartments was carried on two six-wheel steam bogies of plate frame construction. The cylinders were mounted at the rear of each steam unit and the steam chests were above the cylinders, the balanced slide valves being actuated by Walschaerts valve gear, reversing being effected by a hand wheel in the cab. The springs of the coupled wheels in each unit were equalised and all wheels were flanged. The boiler had taper barrels with three rings, the dome being mounted on the back ring. A copper Belpaire firebox was provided and had the usual direct roof stays. The boiler was originally fed by two American lifting type injectors on the firebox front delivering by internal pipes, but in 1912/13 these were altered by the Locomotive Superintendent Mr. P. C. Dewhurst, and mounted on top of the side tanks delivering to a double top feed check box between the bell and the smokebox. Ramsbottom safety valves were fitted. As supplied these locomotives had

rear chimneys but these were soon removed, the rear steam bogie exhaust being taken forward to the smokebox. Unfortunately this arrangement caused excessive draught resulting in spark throwing and for a while shortly after construction No. 32 carried a diamond stack in an attempt to overcome this problem. It was however, unsuccessful and soon removed. Equipment included 11in air compressors, Westinghouse air brakes and a hand brake. All three locomotives were later modified by the addition of a backward extension to the limit of the frames of the rear tank and bunker—No's 30 and 32 being altered after only a year or two's service and No. 31 being modified in 1912. Shortly after construction, No. 32 received steam reversing gear but this was soon removed.

The three Jamaican Kitson Meyer's were, in their original condition, far from satisfactory as confirmed by a letter from the Acting Locomotive, Carriage & Wagon Superintendent Mr P. C. Dewhurst, to Kitson & Co, Leeds, dated 2 February 1918.

"Gentlemen:—

We have here a class of KitsonMeyer tank engines built by you in 1904 (Makers' No's 4252-4).

These engines have always been unsatisfactory, principally on account of the boiler being too small to supply the four (4) high pressure cylinders of 13in × 22in. Two (2) of them were withdrawn by my predecessor here, but owing to the extreme shortage of engines, I intend to resuscitate them and put them back into service.

I am going to convert them to compound taking the H.P. steam from the dome through the back wrapper plate of firebox (if I can manage it) and so to the back engine. From the back engine the exhaust from the H.P. will go to the front engine which will be L.P. Starting valves and intercepting valves will be fitted.

If I can arrange to reverse the trucks, I am going to do so, in order that the exhaust from the L.P. will have a short, easy run out to the blast pipe. I am, of course, altering the cylinders.

Now, these engines have been troublesome with the steam-pipe flexible joints, and in making the conversion I want to fit the latest arrangements. When on the Chilian Transandine Railway some years ago I had experience with flexible steam-pipes on your "Rack & Adhesion" locomotives: the joints there, certainly gave less trouble than did ours here, but I believe you have still further improved them since.

I should, therefore, be obliged if you would send me particulars (and prints, if possible) of your latest steam-pipe arrangements as applied to the Chilian Transandine Engines and any other information you may care to send. . ."

The reply from Kitson & Co. dated 5 March 1918 read as follows:

Dear Sir,

We are interested by the remarks per your letter of 2 February, regarding the relative boiler power of the 2—8—0 and Kitson-Meyer type Locomotives, supplied by us to the Jamaica Government Railway, in the years 1901 and 1904 respectively.

2. With reference to the design of these engines, we wish to mention, as being of interest in consideration of their development, that in each case, a stipulated tractive force had to be developed on a prescribed maximum axle load. So that, after the engine power was attained, there only remained the boiler as the principal item, by which the limiting weight could be

secured.

3. Under these conditions, the Kitson-Meyer Engine was provided with a boiler which was considered by the Railway Companies representatives, to be of such ample power, that it would probably be found, that the cylinder diameter of 13in could be increased after service trials. For this reason, the cylinder barrels were made sufficiently thick, to permit boring out if desired. As the boilers are stated to be small and bad steamers, we should be glad to know the calorific value of the fuel used, prior to the engines being taken out of service.

4. With regard to an increase in the relative power of the boiler over that of the engines, in the Kitson-Meyer Engine, we consider this could be attained in the easiest and quickest manner by inserting liners in the cylinder barrels and cabled this suggestion to you on 22 February as follows:—
"Suggest you try half inch liners in cylinders reducing diameter to twelve-inches. Writing. Kitsons."

5. By this means, new pistons and cylinder covers would only be required, and the engines could be placed in service more quickly than would be the case, if they had to be altered for compounding or superheating, to obtain increased boiler power."

P. C. Dewhurst had inked in the following comment on receipt of Kitson's letter:
"This expedient will give us no more power per lb of coal than before".

The letter continued:

6. In compliance with your request, we enclose herewith print No. 47/18, which shows the method of leading the steam pipes from the Regulators in the Dome to the front and hind bogies respectively, of the "Rack and Adhesion" engines, we built for the Chilian Transandine Railway Co."

Following receipt of the above letter P. C. Dewhurst wrote on 15 April 1918 to the Director of Government Railways, Kingston:
"I send herewith the correspondence with Messrs. Kitson & Co. relative to the Kitson-Meyer Engines . . .

The expedient suggested in their Paragraphs 4 and 5 of reducing the cylinder diameter is not at all good; it will give 23080 lbs R.T.F. for a consumption of steam of 2770.2 cubic feet per mile at 50% cut off and wil reduce coal consumption 14.7% (the same amount as the reduction in power), whilst by compounding the engines on my system we get 22176 lbs R.T.F. (clearance diagram only prevents more power) for a consumption of steam of 1870.3 cubic feet, which will reduce coal consumption by 30%.

Re Paragraph 7, Kitson's have not noticed that the two trucks can be reversed if they are also changed from front to back of the engine thus making unnecessary any alteration of buffers, draw-gear, etc.

There are of course quite a number of calculations and comparisons of ratioes, etc. bearing on this question all of which were taken into consideration before deciding that it was best to compound these engines, and as they are too volumi-nous to be put in a letter, if you wish to have any further points cleared up I would go through them with you personally."

The following extracts are from a letter written from Jamaica by P. C. Dewhurst to Colonel Edwin Kitson-Clarke, dated 14 May 1920 and is informative regarding the JGR Kitson Meyer locomotives. This was kindly supplied by Mr. P. K. Dewhurst.

". . . . First of all I must explain that owing to being so busy putting things ship-shape since my return your letter has remained unanswered for some six weeks, but having now got something like order I am dealing with it.

As I mentioned to you as expected, when I returned I found everything had been interfered with—to such an extent that the new Director of the Railway (whom I had met in England shortly after my Leeds visit and found to be a man who would put things straight) cabled home for my return as early as possible in order to stop the rot—and that was the reason for my rather hurried departure towards the end of December.

Amongst other things a number of details of my design for our new heavy Engines 4—8—0 type were altered, and although, fortunately I arrived in Jamaica in time to save most of them during construction in Canada, yet the interference with the earlier ones had caused considerable trouble and expense and I am now altering same as fast as possible in the Shops.

The third Kitson-Meyer which was almost completed when I left was put into the scrap-heap—on the pretext of a flaw in the tube-plate—but I am of course bringing it into Shops again as soon as other work will allow and when the tube-plate which I order by cable from you is to hand. Efforts have been made to discredit the two already in service but without success,—they are doing well. The coal consumption in 1914-15 when the Engines were still in good average repairs some 12 months prior to their withdrawal from service was 76lbs per Engine mile, for the past three months, January—March, 1920 the average consumption of two altered Engines is 68lbs per mile on the same service but with slightly increased average loads compared to the 1914 loading. As the feed water heating arrangements which I had provided for have not been fitted up during my absence the improved efficiency is evidently entirely due to the modification of the steam and exhaust pipes etc".

P. C. Dewhurst continues: "The excessive flange wear and failure of axles, which was so troublesome on the main-driving wheels of the front truck, has ceased, evidently as result of the reversing of it having had the desired effect of relieving that axle of its double duty of taking care of the principal driving power and curving the engine at the same time. The weight on the respective trucks now only differs by one Ton; this being due to the change in position of the truck centre on the main frames.

The piping has remained untouched since the engines went into service after conversion; the special sliding section of the front engine exhaust-pipe is a great success never having leaked since erection and the engines are working 24 degree curves.

The lower part of the exhaust pipe is also good and the engines are excellent steamers, although prior to alteration that was, of course, their chief failing.

Larger Chimneys have been fitted and two 9 inch Air-pumps in place of the one 11 inch pump originally fitted.

Note:—The rear tank was enlarged and the rear exhaust originally carried to the smokebox (when a "double" Blast-pipe was fitted) some years ago.

Engine 31 now has cylinders 12in × 22in and Engine 30 has cylinders 12in × 22in".

I may say that the reason I held up the compounding of the third Engine of this type was due to the fact that I had reason to suppose that it would not be made a success during my absence and eventually it is proved that I was correct; the question as to whether I shall compound the third one, when I am able to

Jamaica Railway Kitson Meyer Type 1 0−6+6−0T No. 30, Kitson Works No. 4252/1904.   *D. Binns collection.*

No. 30 as modified by P. C. Dewhurst to Type 3 configuration by the turning round of the front truck. Note also that the rear tank has been extended backwards to the limit of the frame and the rear chimney has been removed. The front end footplate has also been altered and an air compressor fitted; also electric lighting. Observe also the gap between the leading truck and front of the firebox, created by reversing the front bogie.   *P. C. Dewhurst via P. K. Dewhurst.*

JGR No. 30 from the opposite side showing the extra Westinghouse air compressor and the injector mounted on top of the side tanks delivering to a double top feed clackbox between the bell and the smokebox. Note the dumb buffer with central buckeye coupling.   *P. C. Dewhurst via P. K. Dewhurst.*

tackle it, is not yet decided for the reason that the first two are really doing very well as so far altered.

The up-shot of the whole matter is that for these and other similar reasons, and as the new Director has decided to go elsewhere, I have formally notified the Local Government and the Colonial Office at home that I desire a transfer and I am making my plans for the future without regard to Jamaica . . ".

In correspondence to me Mr. P. K. Dewhurst (P.C's son) stated:

"I remember P.C.D. told me that the principal problem with the KMs as built was their tendency to run out of steam on the long climbs. In fact that weakness caused all three to be taken out of service in 1916. Tyre wear was a bad but secondary problem.

In 1919 he resuscitated the KMs (as he called it) and redesigned the steam pipes etc. and brought the rear exhaust to aid the draft. He realized that the boilers would have to be forced in order to produce enough steam to operate at the late cut offs necessary to haul the loads up the gradients. He sent drawings of his modifications to Col. Clarke to use as he thought fit in the future. This was the style of their relationship. In 1922, Edwin Kitson Clarke delivered P. C.

Dewhurst's paper to the Institute of Mechanical Engineers (for which he received a gold medal which I now have) because the J.G.R. would not allow P.C.D. time off to deliver it himself.

The most obvious and visual change was the turning round of the leading bogie which did in fact reduce flange wear. No's 30 and 31 were modified early in 1919, prior to P.C.D's visit to England. No. 32 followed after Kitsons had sent a new tube plate; but not until after 1921 I think".

I knew that PCD had considered compounding these KM's at one time (I have copies of his correspondence with Baldwins on the subject of Mallets and "starting valves") I also know that although this was theoretically the best solution it was not carried out because of the cost in relation to the remaining useful life of the locos.

After that period the new type 4−8−0's took over all the heavy main line gradient work .

The turning round of the leading steam bogie converted the three Kitson Meyer's to pseudo-type three, i.e. with cylinders at the outer ends of each steam bogie (as delivered there was virtually no overhang at the front end but as altered the cylinders were ahead of the smokebox. As delivered the steam bogies were pivoted 10in in front of the middle coupled axles

JGR No. 31 with cylinders at the rear of each bogie but minus rear chimney.
JGR No. 31 as rebuilt by P. C. Dewhurst.

P. C. Dewhurst via P. K. Dewhurst.
P. C. Dewhurst via P. K. Dewhurst.

on each unit, but when the front bogie was reversed the pivot point was 10in behind the middle coupled axle on the leading unit. In service these units proved heavy on coal and repair when compared to conventional motive power doing the same work and no further Kitson Meyer locomotives were purchased.

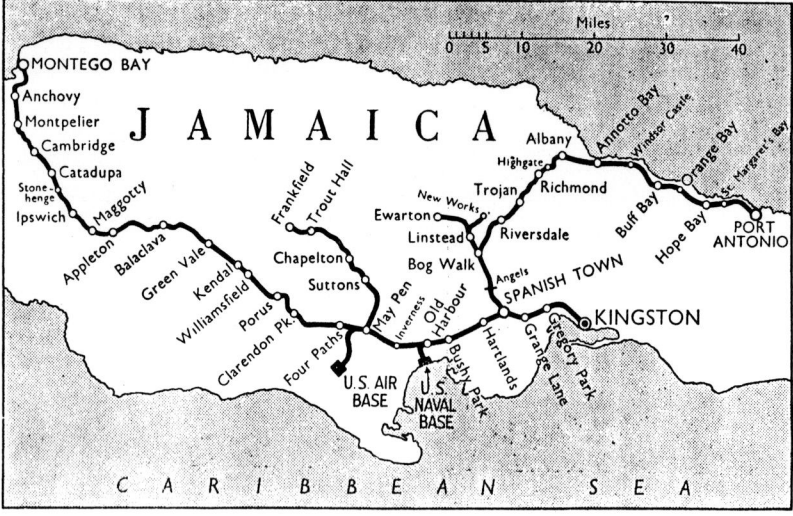

Map courtesy "The Railway Magazine". November/December 1945.

Kitson Meyer locomotives supplied to Jamaican Government Railways

0−6+6−0T. Built by Kitson & Company 1904. Three locomotives supplied. Works No's 4252/3/4. Became JGR Road No's 30, 31 and 32. Type 1 simple tank locomotives with the cylinders mounted at the rear of each steam unit. Steam trials 27/7, 3/10 and 15/11/1904.

Railway	Jamaica
Gauge	4ft 8½in
Wheels	0−6+6−0T
Maker	Kitson
Works No.	4252/3/4
Year	1904
Cylinder Position	rear
Cylinders−inches	13 × 22
Boiler Pressure−lbs/sq in	180
Heating Surface:	
Firebox−sq ft	130
Tubes−sq ft	1328
Total−sq ft	1458
Superheater−sq ft	None
Total HS−sq ft	1458
Grate area−sq ft	26
Driving wheel−diameter	3ft 6in
Other wheels−diameter	None
Rigid wheelbase−first group	7ft 9in
Rigid wheelbase−second group	7ft 9in
Total wheelbase	29ft 9in
Water−galls	2500†
Coal−tons	4
Weight empty−tons/cwt	60 0
Weight in W.O.−tons/cwt	80 15†
Weight Adhesive−tons/cwt	80 15
Tractive effort−lbs	23,895 (75%)
Height−overall	13ft 9in
Width − ”	10ft 0in
Length− ”	40ft 5in
Hauling capacity on straight level track @ 8−10mph	2877 tons
On 1 in 100 @ 8−10mph	698 tons
On 1 in 75 @ 8−10mph	545 tons
On 1 in 50 @ 8−10mph	367 tons
On 1 in 25 @ 8−10mph	162 tons

† Water capacity later altered to 2850 gallons. Weight then increased to 82 tons 4 cwt.

No. 32 as modified by P. C. Dewhurst to pseudo Type 3 with cylinders at the outer ends. A generator is mounted behind the dome to supply electricity for the head and back-up lights.
P. C. Dewhurst via P. K. Dewhurst

SPAIN

Three large 5ft 5¹³⁄₁₆in gauge, 2−8+8−0T Kitson Meyer articulated locomotives, were supplied by Kitson & Co. in 1908 to the English owned and managed Great Southern Railway of Spain, for use on both passenger and heavy mineral trains. The Company had been formed in 1885 to build a trunk line from the Atlantic to the Mediterranean and its 104 miles of track commenced at Granada, and terminated at Murcia, where it joined a line to Madrid. The Kitson Meyers were for use on the lightly tracked routes of the Compània de los Ferrocarriles de Lorca a Baza y Aguilas (the Lorca−Baza−Aguilas Railway. The Great Southern of Spain, together with all other Spanish broad gauge railways, amalgamated in 1941 to form Red Nacional de los Ferrocarriles Espanoles (RENFE). At some time before this the 2−8+8−0T were "repaired" by Babcock & Wilcox Espanola, but this amounted to a virtual rebuilding and several differences will be observed in the accompanying photographs. They finished their days on the Córdoba−Peñarroya section being withdrawn in 1953 and broken up soon after.

Kitson Meyer locomotives supplied to the Great Southern Railway of Spain

2−8+8−0T. Built by Kitson & Company 1908. Three locomotives supplied. Works No's 4580/1/2. Type 3 simple tank locomotives with the cylinders mounted at the outer ends of each bogie. Road No's 50−52. Steam trials 25/8, 4/9 and 24/9/1908.

Railway	Gt S (Spain)
Gauge	5ft 5¹³⁄₁₆in
Wheels	2−8+8−0T
Maker	Kitson
Works No.	4580/1/2
Year	1908
Cylinder Position	outer
Cylinders−inches	14¾ × 24
Boiler Pressure−lbs/sq in	180
Heating Surface:	
Firebox−sq ft	132
Tubes−sq ft	1716
Total−sq ft	1848
Superheater−sq ft	None
Total HS−sq ft	1848
Grate area−sq ft	34.6
Driving wheel−diameter	4ft 0in
Other wheels−diameter	2ft 9in
Rigid wheelbase−first group	15ft 0in
Rigid wheelbase−second group	15ft 0in
Total wheelbase	49ft 3in
Water−galls	2300
Coal−tons	2½
Weight empty−tons/cwt	101 0
Weight in W.O.−tons/cwt	109 0
Weight Adhesive−tons/cwt	90 12
Tractive effort−lbs	29,322 (75%)
Height−overall	14ft 10in
Width − "	9ft 9¼in
Length− "	60ft 10in

Locomotoras-ténderes núms. 180-0401/180-0403

Procedencia: F. C. Lorca a Baza y Aguilas (núms. 50-52).
Construcción: Kitson & Cía.—Año 1908.

DIAGRAMA

MAQUINA-TENDER

Cilindros: Diámetro interior................. d= 374 m/m.
Carrera del émbolo................. L= 609 m/m.
Distribución plana Stephenson.
Ruedas: Diámetro de las motoras........... D=1.219 m/m.
Caldera: Timbre........................ p=15 kgs./cm².
Diámetro interior del cuerpo cilíndrico. 1.621 m/m.
Longitud entre placas tubulares...... 4.705 m/m.
Tubos: Diámetro exterior................. 50,7 m/m.
Número...................... 218
Capacidad: Agua..................... 12,712 m³.
Carbón.................. 6.000 Kgs.

Superficie de { Hogar......................... 13 m².
calefacción: { Tubos........................ 163,6 m².
Total..................... 176,6 m².
Superficie de la rejilla.................. 3,21 m².
Peso: Locomotora vacía................. 82.000 Kgs.
Locomotora en servicio........... 102.615 Kgs.
Adherente.................. 92.049 Kgs.
Por metro lineal de locomotora....... 5.443 Kgs.
Esfuerzo de tracción $F = \dfrac{0{,}65 \, p \, d^2 \, L}{D}$ .... 13.750 Kgs.
Potencia normal indicada.............. 1.345 C.V.
Alumbrado por petróleo.

Above: The first of three 2–8+8–0 Kitson Meyer locomotives supplied to Spain in 1908. These were the first Type 3 and the only ones to have bogies with different wheel arrangements. The circular number plate reads (round the outer edge): "Great Southern of Spain Railway, Ferrocarril de Lorca a Baza", (inside ring): "Kitson & Co. Ltd., Leeds, No. 4580, 1908", with the running number "50" in the centre.
D. Binns collection.

The three Great Southern of Spain Railway 2–8+8–0T were eventually "repaired" by the Sociedad Espanola de Constructiones Babcock & Wilcox, but this amounted to a virtual rebuilding. The most obvious difference is in the built up RENFE bunker.

Works No's 4580/1/2 of 1908

←

				11.40	11.85	11.90	11.90	**Loaded 100.38 tons**
								" 89.98 Adhesive
				10.90	10.95	11.45	11.30	**Unloaded 94.60**
10.40	11.25	11.35	10.18	10.15				
9.75	10.25	10.35	10.00	9.65				
Tare 80.75								

Great Southern of Spain Railway 2−8+8−0T No. 50 heads a mineral train. Location and year unknown.
J. V. Coves Navarro collection.

The end of one of the Great Southern of Spain Railway 2−8+8−0T at Cordoba Arcadilla on 18 October 1952.
G. Reder.

No. 50 with a mineral train on the GS of S Railway.
A. E. Durrant collection.

Manila

The 3ft 6in gauge Manila Railroad opened its first section between Manila and Dagupan, a distance of 121½ miles, on 24 November 1892. At that time the Philippine Islands were a Spanish colony and in 1898 following events in Cuba, war broke out between Spain and America resulting in American occupation of the Philippines. Eventually when American sovereignty was established, the British Manila Railway Company formed a subsidiary based in New Jersey USA, to operate the Manila Railroad. In 1917 the Company passed to the Philippine Government.

Four Kitson Meyer 2−6+6−2T were supplied in 1913 to the specifications of Mr. R. D. Deacon, Locomotive Superintendent of the Manila Railway and built by Kitson & Company under the supervision of the consulting engineers D. M. Fox & Sons, of London. The locomotives were type 3 simple tanks with the cylinders at the outer ends of each steam unit and they had Robinson superheaters, piston valves, full length side tanks, Wakefield mechanical lubricators and rear chimneys. On which part of the system the Kitson Meyer's were employed is not known, but at some time before the First World War the Company contemplated building a mountain section from Damortis to the hill station of Baguio and in fact took delivery of six 8-coupled rack/adhesion locomotives for this route which was never completed. Is it possible that the Kitson Meyer locomotives were purchased for this line?

Kitson Meyer locomotives supplied to the Manila Railroad

2−6+6−2T. Built by Kitson & Company 1913. Four locomotives supplied. Works No's 4972/3/4/5. Road No's 140−143 (161−164 on 4/1914). Type 3 simple tank locomotives with cylinders at the outer ends of each steam bogie. Steam trials 22/8, 29/8, 16/9 and 3/12/1913. Withdrawal dates unknown.

Railway	Manila
Gauge	3ft 6in
Wheels	2−6+6−2T
Maker	Kitson
Works No.	4972−5
Year	1913
Cylinder Position	outer
Cylinders−inches	16 × 20
Boiler Pressure−lbs/sq in	160
Heating Surface:	
Firebox−sq ft	136.5
Tubes−sq ft	1612.2
Total−sq ft	1748.7
Superheater−sq ft	352
Total HS−sq ft	2100.7
Grate area−sq ft	35
Driving wheel−diameter	3ft 3in
Other wheels−diameter	2ft 0½in
Rigid wheelbase−first group	7ft 0in
Rigid wheelbase−second group	7ft 0in
Total wheelbase	39ft 7in
Water−galls	3000
Coal−tons	4
Weight empty−tons/cwt	69 13
Weight in W.O.−tons/cwt	92 8
Weight Adhesive−tons/cwt	71 18
Tractive effort−lbs	31,440 (75%)
Height−overall	12ft 10in
Width − ”	8ft 9in
Length− ”	48ft 4in
Hauling capacity on straight level track @ 8−10mph	3491 tons
On 1 in 100 @ 8−10mph	900 tons
On 1 in 75 @ 8−10mph	707 tons
On 1 in 50 @ 8−10mph	483 tons
On 1 in 25 @ 8−10mph	219 tons

"KITSON MEYER" LOCOMOTIVE FOR THE MANILA RAILWAYS.

TYPE ... 2.6.0.6.2. GAUGE ... 3 FT. 6 IN. = 1066·8 M/M.

Manila Railways 2−6+6−2T Type 3 Kitson Meyer No. 143 was later renumbered 164 and carried Works No. 4975/1913. Few photographs seem to exist showing these locomotives and the one reproduced has been taken from the Kitson & Co. 1923 catalogue.

D. Binns collection.

INDIA

Two particularly fine Kitson Meyer locomotives were built by Kitson & Co. in 1928 for the North Western Railway—owners of the Kalka—Simla line. The 2ft 6in gauge Kalka Simla Railway extended from the 5ft 6in gauge railhead of the East India Railway at Kalka (2,143ft above sea level) and proceeded to Simla, the summer headquarters of the Indian Government, located in the foothills of the Himalaya mountains, the line terminating at Simla goods station (6,870ft). The railway 59.9 miles long opened in 1903, being purchased on 1 January 1906 by the Indian Government and worked by the North Western State Railway from 1 January 1907. The track climbed continuously for the first 20 or so miles at 1 in 33 uncompensated for curvature, then fell and climbed again at grades varying between 1 in 33 and 1 in 40. Curves were extremely sharp and in all there were over 1,000 including 220 of 120ft radius and one of only 110ft. All curves of 140ft radius and less had check rails and were super-elevated up to 2½in. This meant that bogie vehicles and of course the two Kitson Meyer locomotives had one bogie tilted on its east side by 2½in and the other tilted 2½in on its west side as they passed through reverse curves. Further complications of the Kalka Simla Railway were 103 tunnels including one ¾ mile long at Barogh. Originally laid with 41lb/yd rail the route was later re-laid with 62lb rail on wood sleepers ballasted with stone.

The railway played an important part in transporting the Indian Government up and down between its summer and winter capitals and until the late 1920s carried heavy all year round traffic. Trains also had to contend with regular snow storms during the winter months and before the arrival of the Kitson Meyer locomotives motive power was provided by 2−6−2T hauling 85 tons unassisted.

By 1928 road competition had become a serious problem and the North Western Railway decided to introduce 160 ton train loads hauled by articulated locomotives working at 10mph. An approach was made to Rendel, Palmer & Tritton of Westminster, Consulting Engineers to the Indian State Railways, who after much deliberation ordered two 2−6−2+2−6−2T Kitson Meyer locomotives. Kitson & Company supplied both during 1928 and these were set to work on goods trains in the following year. These were exceptionally fine units and were of necessity extremely flexible and in order to ensure success under the difficult operating conditions of the Kalka—Simla, incorporated certain design modifications. The contour of the tyres was modified to a very full shape to combat excessive wear expected on the severe curvature of this railway. On both steam bogies the spring gear was compensated in two groups, the leading and trailing wheels were provided with vertical movement combined with the spring compensation. Outside journals were provided for the leading pony trucks at both ends and inside journals to all other wheels. In these locomotives exhaust from the rear steam bogie was carried to the blast pipe through a rectangular pipe beneath the right hand side tank and a further pipe carried steam from the header to the rear unit. Single brake blocks were provided on all coupled wheels operated by vacuum brake and an automatic water spray was incorporated to cool down the blocks following lengthy applications. Reports indicate that the Kitson Meyer locomotives had a tendency to slip when passing through reverse curves but otherwise were highly satisfactory.

The Engineer for 2 November 1928 had this to say:

"Some time ago we had, through the courtesy of the builders, an opportunity of seeing the first of these engines put through shop tests. A short length of line bent into an S with considerable super-elevation had been laid down, and the engine was run backwards and forwards under her own steam. Of course, speed was out of the question, and the test was really one of flexibility. The total wheel base of the engine is 44ft 10in long, long enough to extend over the greater part of the double curvature. Hence whilst the forward engine was turned, let us say, to the right and twisted by the corresponding super-elevation, the after engine was turned and twisted in the opposite direction; the maximum twist, however, was not obtained with both ends at once. It is difficult to give any just impression of the extraordinary antics which, in these circumstances, the locomotive appeared to be performing, but the little illustration ★ may give some idea of it. It must be remembered, for the illustration does not show it very clearly, that, since the boiler is rigid and the forward end of the 'bogie'

★ *The little illustration is not of good enough quality to reproduce.*

North Western Railway (India) 2−6−2+2−6−2T No. 181 Kitson & Co. Works No. 5414/1928 was photographed on the Kalka-Simla section (photographer unknown). D. Binns collection.

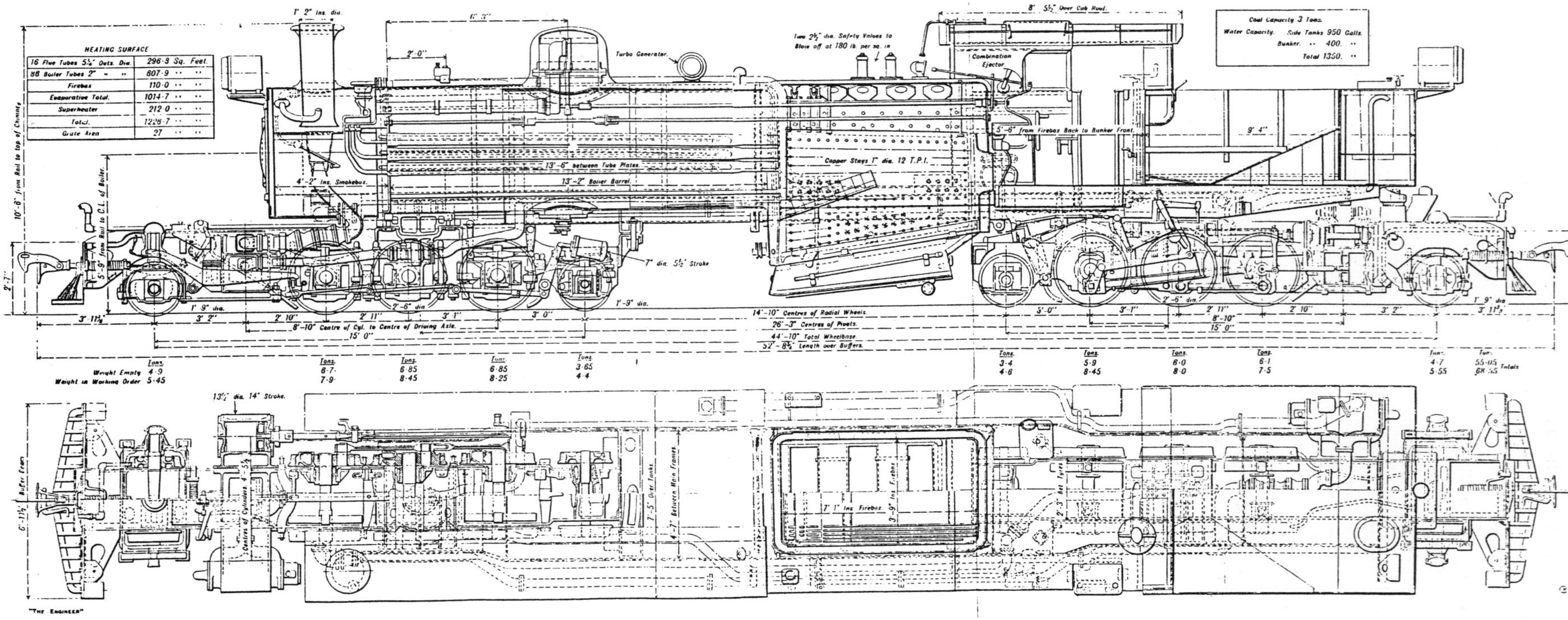

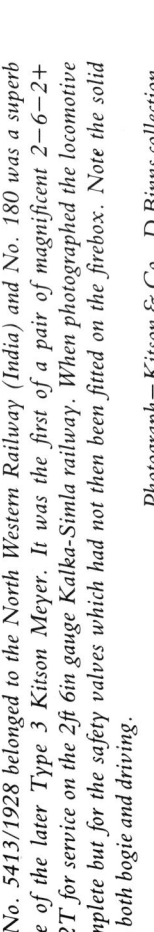

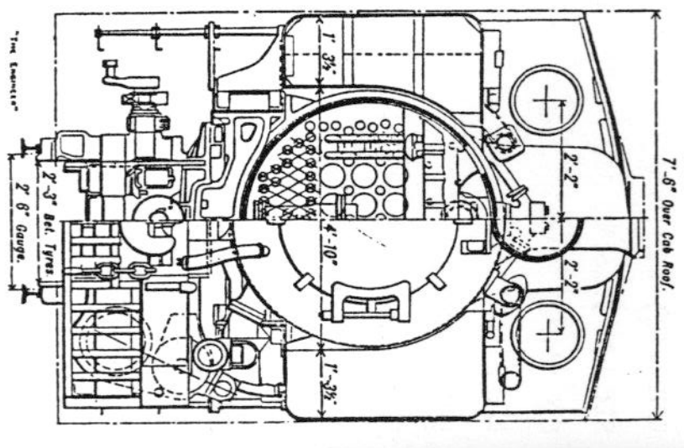

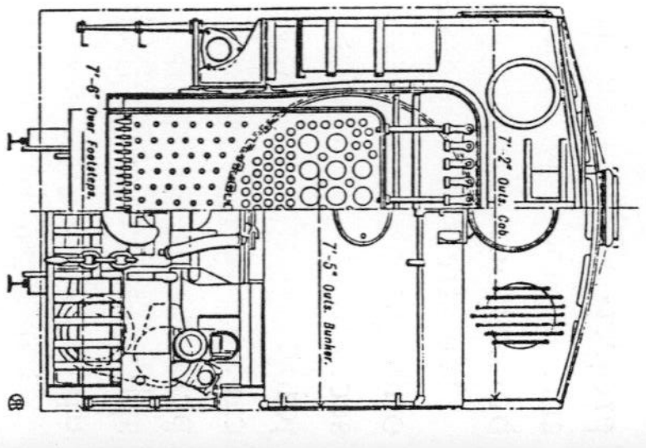

Works No. 5413/1928 belonged to the North Western Railway (India) and No. 180 was a superb example of the later Type 3 Kitson Meyer. It was the first of a pair of magnificent 2–6–2+ 2–6–2T for service on the 2ft 6in gauge Kalka-Simla railway. When photographed the locomotive was complete but for the safety valves which had not then been fitted on the firebox. Note the solid wheels, both bogie and driving.

Photograph– Kitson & Co., D Binns collection.
Drawings– "The Engineer", 2 November 1928.

projects 12ft or 13ft in front of its pivot, the 'bogie' twists and inclines by an astonishing amount relatively to the boiler front.

We stress this extraordinary flexibility because it is the outstanding characteristic of the Kitson-Meyer system. It is secured, as is well known, by the use of double spherical joints in the steam and exhaust pipes combined with a careful arrangement of clearances, a type of axle-box which permits considerable movement in the hornblocks, and a spherical bolster which gives the 'bogies' freedom to take up any necessary angle.

The arrangement of the flexible joints in the steam and exhaust pipes can be seen in the forward engine. In the trailing engine a modification had to be made, since the steam has to be taken to the rear from the header and the exhaust has to be carried forward to the blast nozzle. For this purpose pipes which are shown in dotted lines are provided. The exhaust pipe, which can be seen just under the tank in the side view, is rectangular in section.

A word must be said about the driving wheels. It will be noticed that none of them are blind, they all have flanged tyres, but owing to the heavy wear caused by the sharpness of the curves, the contour of the flanges is much fuller than usual, the familiar radius at the top being reduced to little more than a rounding off of the corners.

A careful study of the drawing will reveal several other interesting points; it will, for example, be seen that, whereas the two end pairs of wheels are of the pivoted type, the two inner pairs are carried in radial axle boxes, and that single brake blocks, applied to all the drivers, are employed. A little turbo-generator for the headlight will be noticed on the boiler top, and it will be seen that the sand boxes are placed low down, the end pairs on the top of the frames and the inner pairs between the inner carrying wheels and the adjacent drivers; and automatic water spray for cooling the brake blocks when applied is also fitted.

Both engines have now been re-erected in India, and we believe one at least is already on the road, but it is too early yet to get reports of its behaviour. We venture, from our own inspection, to congratulate the makers and the consulting engineers on the production of a really remarkable engine, remarkable for the train loads—exclusive of the locomotive—it is expected to haul, upwards of 160 tons, on a road which is about as difficult as possible to imagine and up steep gradients at a speed of over ten miles per hour, and for a flexibility which has to be seen to be believed."

After working for a couple of years or so on the Kalka—Simla section, the Kitson Meyer's were transferred to the Kangra Valley Railway—the reason being a decline in traffic resulting in these locomotives being left in steam for up to eight hours until suitable loads could be made up. Obviously the cost of this enforced idleness was a set back against other economies gained by increased train loads.

The Kangra Valley Railway came into being during the 1920s following a decision by the Punjab Government to construct an important hydro-electric project. The Uhl river in Mandi State was dammed above Jogindernagar, the State Capital, and water was piped down to a turbo-generating station capable of sufficient output to supply the entire Punjab with electricity. To enable heavy machinery and equipment to be transported from the broad gauge railhead at Pathankot, to Jogindernagar 103 miles away, the 2ft 6in gauge Kangra Valley Railway was built. In addition to carrying machinery and equipment the line also served the tea gardens of the Kangra district of the Punjab. The KVR was laid with 40lb/yd rail, easy curves and 1 in 40 grades between Pathankot and Baijnath Paprola 89 miles away but the last 14 miles to Jogindernagar was built through difficult country with a 1 in 25 ruling grade. Until the arrival of the Kitson Meyer locomotives 140 ton trains were hauled by 2—8—2 tender types as far as Baijnath Paprola where part of the load was dropped, the remaining 14 miles being worked by three locomotives (a 2—8—2 with two 2—6—2 bankers). The two Kitson Meyer's were employed as bankers with one 2—6—2 type as train engine between Baijnath

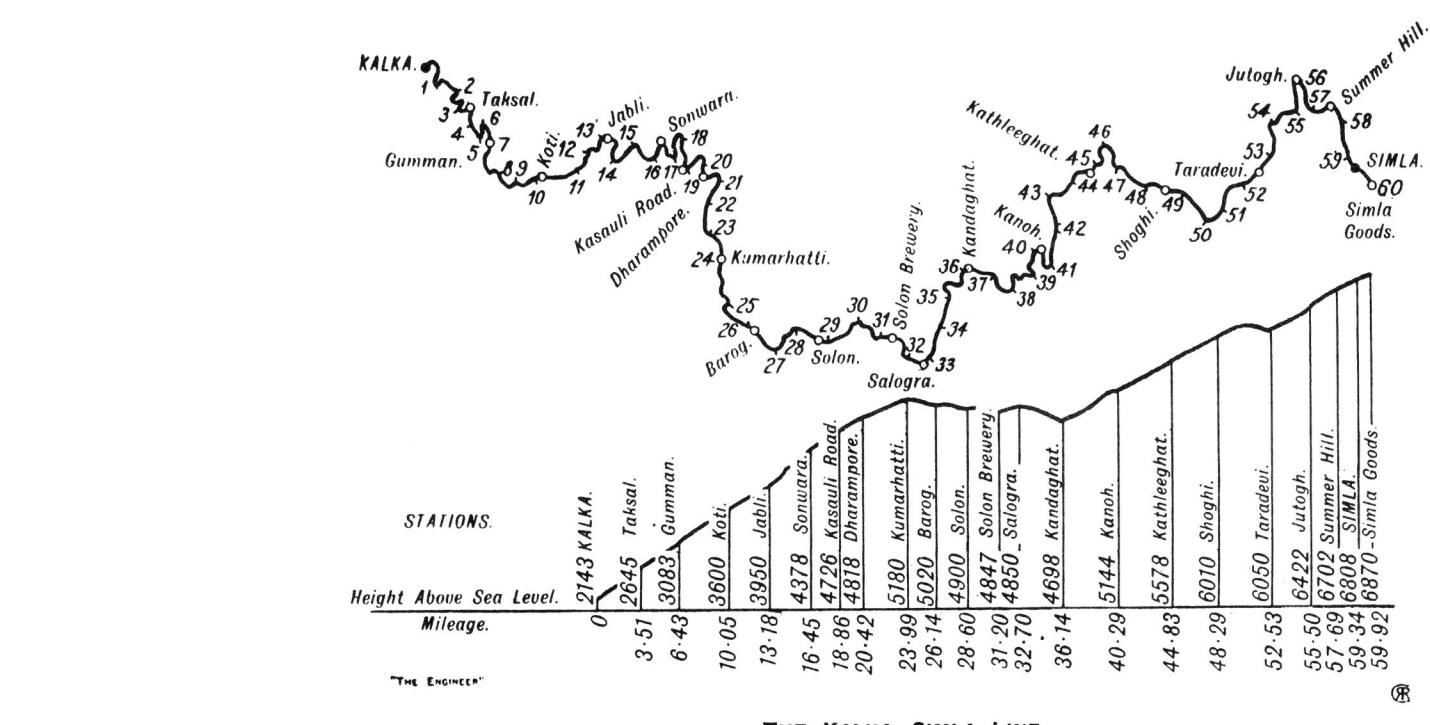

THE KALKA-SIMLA LINE

The second 2−6+6−2T carried North Western Railway No. 181 and this locomotive was photographed at work in India. *M. A. Harrison.*

Paprola and Jogindernagar on which duties the 2−6−2+2−6−2T proved highly reliable and excellent steamers when worked hard. Both remained on the Kangra Valley until 1939 when the upper section was closed to passenger traffic, being used as a siding for the hydro-electric power station. The Kitson Meyer's being non-standard were despatched to the railway works at Mogholpura for disposal. Whether due to war needs they were kept in service is not known but there was only one other 2ft 6in gauge line to which they might have gone—the 174 mile Hindubogh branch near Quetta. This line was regularly worked by G/S and ZE class 2−8−2s and no surviving records show the Kitson Meyer's as being used there-on. The 2ft 6in gauge Zhob Valley Railway can be discounted due to a 6 ton maximum axle loading.

Mr. A. E. Durrant advises that the two Kitson Meyer locomotives were transferred to Pakistan and converted to 1 metre gauge apparently about 1951. Both were dumped near Hyderabad (Sind) in the 1960s but no photographs appear to exist. Exactly when they were broken up is not known.

Kitson Meyer locomotives supplied to the North Western Railway (India)

2−6−2+2−6−2T. Built by Kitson & Company 1928. Introduced on NWR 1929. Two locomotives supplied. Works No's 5413/4. Became NWR Road No's 180 and 181. Class T/s. Steam trials 28/6 and 17/7/1928. Type 3 simple tank locomotives with the cylinders mounted at the outer ends of each steam bogie.

Railway	NWR
Gauge	2ft 6in
Wheels	2−6−2+2−6−2T
Maker	Kitson
Works No.	5413/4
Year	1928
Cylinder Position	outer
Cylinders−inches	13½ × 14
Boiler Pressure−lbs/sq in	180
Heating Surface:	
Firebox−sq ft	110
Tubes−sq ft	904.7
Total−sq ft	1014.7
Superheater−sq ft	212
Total HS−sq ft	1226.7
Grate area−sq ft	27
Driving wheel−diameter	2ft 6in
Other wheels−diameter	1ft 9in
Rigid wheelbase−first group	6ft 0in
Rigid wheelbase−second group	6ft 0in
Total wheelbase	44ft 10in
Water−galls	1350ø
Coal−tons	3
Weight empty−tons/cwt	
Weight in W.O.−tons/cwt	68 11
Weight Adhesive−tons/cwt	48 11
Tractive effort−lbs	26,025 (85%)
Height−overall	10ft 8in
Width − "	7ft 5in
Length− "	52ft 8¾in

ø Side tank 950 galls., bunker tank 400 galls.

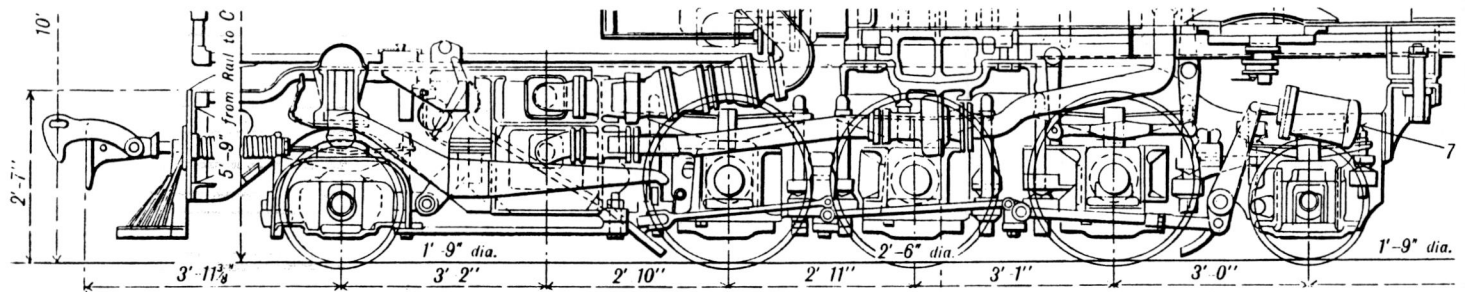

KITSON ARTICULATED CONSTANT ADHESION LOCOMOTIVE

KITSON & CO. LIMITED.

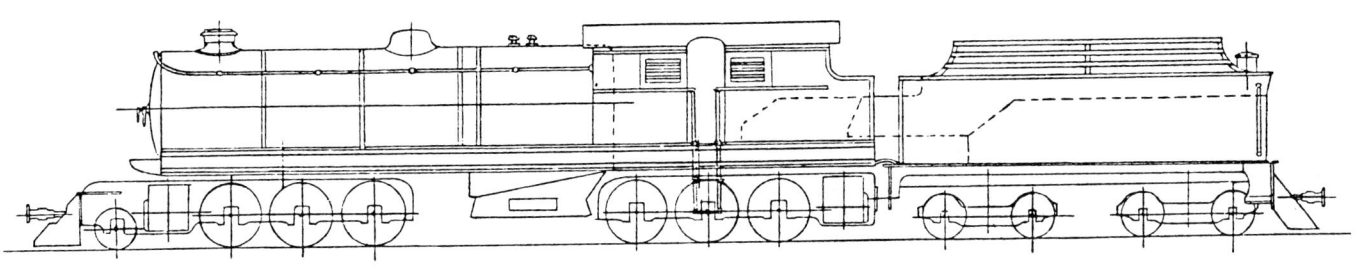

DESIGN FOR A "KITSON ARTICULATED CONSTANT ADHESION ENGINE."

TYPE ... 2.6.0. + 0.6.0. WITH DOUBLE BOGIE TENDER.   GAUGE ... 1 METRE.

ENGINE.

4 CYLINDERS:—Diameter	14 in.	= 355·6 m/m.
Stroke	20 in.	= 508·0 m/m.
HEATING SURFACE:—Small Tubes ...	1013 sq. ft.	= 94·1 M².
Large ,,	366 ,,	= 34·0 M².
Firebox	140 ,,	= 13·0 M².
Total Evaporating Surface	1519 ,,	= 141·1 M².
Superheater ...	285 ,,	= 26·5 M².
TOTAL ...	1804 ,,	= 167·6 M².
FIREGRATE AREA	33 ,,	= 3·06 M².
WORKING PRESSURE	160 lbs. per sq. in.	= 11·25 Kilos per c/m².

WHEELS:—Coupled, Diameter	3 ft. 3 in.	= 993·5 m/m.
WHEEL-BASE:—Rigid	7 ft. 6 in.	= 2285·6 m/m.
Total (Engine)	34 ft. 9 in.	= 10591·0 m/m.
Engine and Tender ...	58 ft. 7 in.	= 17856·0 m/m.
WEIGHT:—In Working Order	... 62 tons	= 62994 Kilos.
On Coupled Wheels	56·25 ,,	= 57152 ,,

TENDER.

TANK CAPACITY	2,140 gallons	= 9728 litres.
FUEL CAPACITY	... 11 tons	= 11·17 tonnes.
WEIGHT, Full	... 36 tons	= 36577 Kilos.
WHEELS, Diameter	2 ft. 4½ in.	= 724 m/m.

TRACTIVE FORCE AT 75 % OF BOILER PRESSURE ... 24120 lbs. =

LOAD EXCLUDING WEIGHT OF LOCOMOTIVE

	Level	1 in 100	1 in 75	1 in 50	1 in 25	SPEED.
TONS OF 2,240 LBS. ...	2646	663	516	344	142	10 miles per hour on straight line.
TONNES ...	2688	673	524	349	144	16·0 km. par heure sur alignement droit.

In their 1923 catalogue Kitson & Co. illustrated a "Kitson Articulated Constant Adhesion Engine" —a 2−6+6−0 with cylinders at the outer ends of each steam bogie. A separate tender was an essential feature to maintain constant adhesion and the adhesive weight was 56¼ tons out of a total engine only weight of 62 tons. The separate tender weighed 36 tons in working order. A patent application had been made on 31 August 1922 by Lieut.-Colonel Edwin Kitson Clark, "of Airedale Foundry, Leeds, in the County of York" and Thomas Haslehurst Brocklebank, the patent being granted in the following year. The specification tells us that the object of the present invention was to provide an improved locomotive of the articulated type. As in the normal Kitson Meyer, the boiler, cab and rear bunker was carried on a girder frame, this being mounted on two pivoted engine driven bogies. A separate tender carried fuel and water supplies—virtually an up-dated version of the unsuccessful type 2 Kitson Meyer of which 4

units were built for South Africa in 1903/4. An improvement in the patent over the type 2 locomotives was that fuel was delivered by a chute from the tender to a receiving bunker inside the cab at the "customary standard distance from the fire-hole" . . . The width of the coal receiver in the cab and the chute from the tender was such that a gangway was provided on each side to enable the fireman to readily move back to the tender and replenish the coal receiver which carried the immediate supply. The coal shute did not rest on the coal receiver and the engine only experienced a variation in load within the limits of the capacity of the receiver, which was relatively small. "In this way a locomotive is obtained in which the engine proper is completely articulated and substantially uneffected by variation in the tender load so that the engine adhesion to the rails is constant."

The drawing reproduced and the dimensions are taken from a Kitson & Co. 1923 catalogue. No actual locomotives were built to this design.

MIDLAND RAILWAY (ENGLAND) KITSON MEYER PROPOSAL.

About 1910/11 the Midland Railway (England) prepared two designs based on the Kitson Meyer form of articulation – both were 2–6+6–2T with the cylinders at the outer ends of each unit and intended for banking on the Lickey incline. The principle difference between the two was in the water capacity, the first carrying 2000 gallons and the later proposal 1000 gallons in low side tanks, presumably lowered to provide better visibility for the engine-men when buffering up to trains. At 75% boiler pressure the tractive effort would have been just short of 40,000lbs. Both these proposed banking locomotives are illustrated by the diagrams – No. DS1677 being the earliest and DS1703 the later one. These came to nothing and were superseded by a further unfulfilled scheme for a rigid frame 0–6–6–0 with 4 cylinders – two at each outer end.

PROPOSED BANKING ENGINE
2–6–6–2
DS–1677

MIDLAND RAILWAY
LOCOMOTIVE DRAWING OFFICE
DERBY

PROPOSED BANKING ENGINE
2–6–6–2
DS–1703

MIDLAND RAILWAY
LOCOMOTIVE DRAWING OFFICE
DERBY

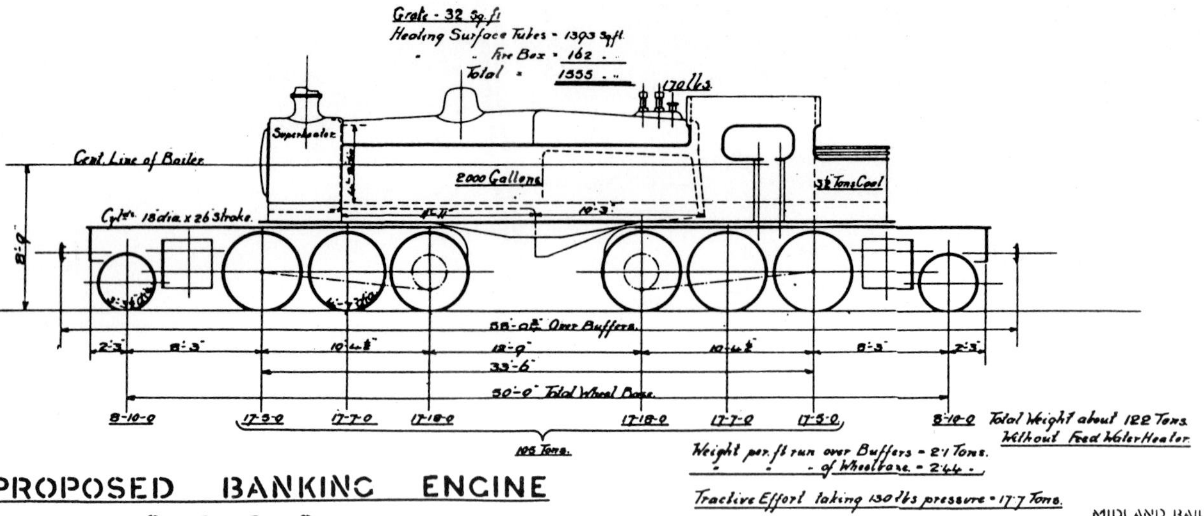

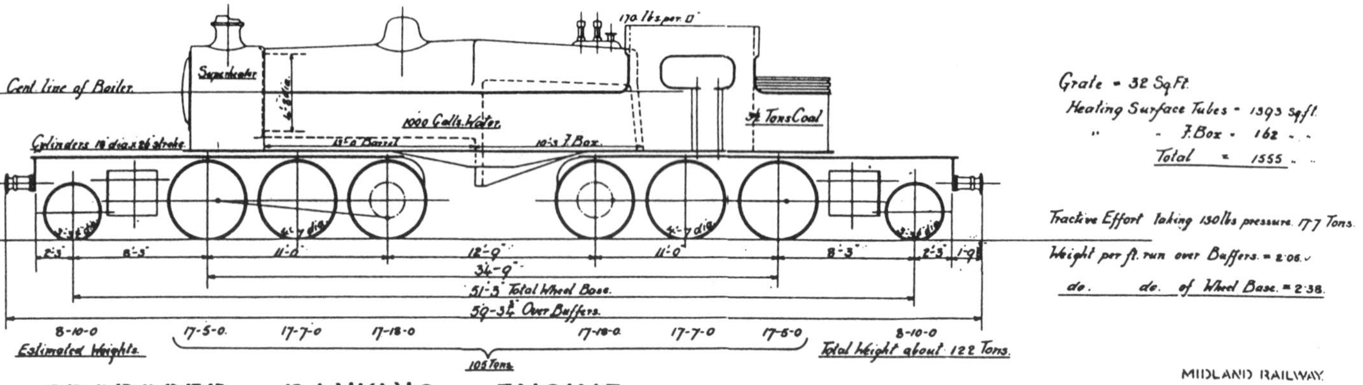

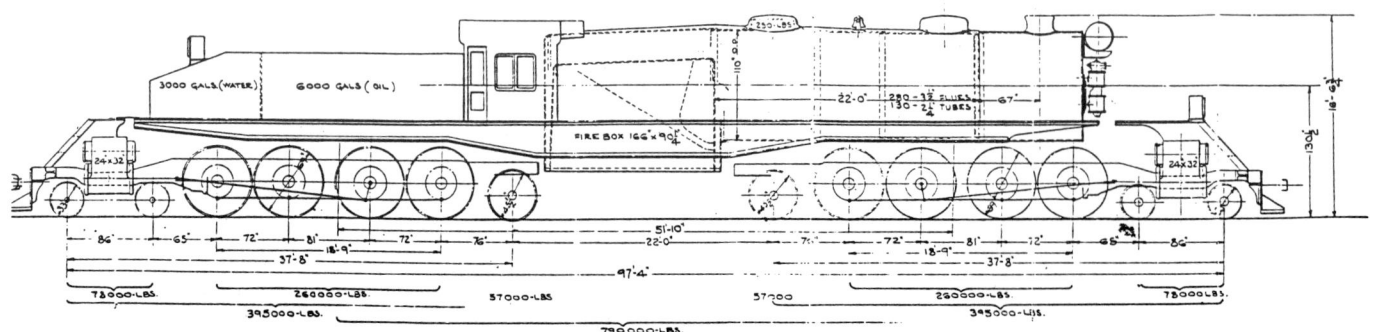

SOUTHERN PACIFIC RAILROAD (USA)

In June 1930 the American Locomotive Company prepared preliminary drawing (S93760) for the Southern Pacific Railroad (USA) for a hugh 352½ ton Kitson Meyer type 4–8–2+2–8–4 with a separate 12 wheel tender. Designated "Fairlie type", the design did not have a leading tank as in the Modified Fairlie,

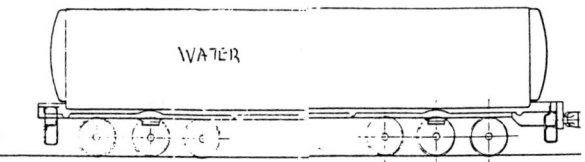